INTRODUCTION

Math plays a vital part in everyone's life every day. Therefore, it is crucial that students develop valuable math skills.

Everyday Problem Solving has been designed to accomplish two goals—to make learning math skills an exciting experience for students and to make teaching math skills equally as exciting for teachers. This book is filled with stimulating games and activities that enable students to see how meaningful and relevant math is in their lives. Skills covered in this book include prime and composite numbers, prime factorization, probability, patterning, addition of fractions with like and unlike denominators, solving equations, exponents, and many more.

The problem-solving games and activities featured in this book are perfect to use to help students learn math skills in a variety of ways. Some activities can be completed individually, some as a class, and others have been adapted for pairs or teams of students.

Each activity includes a list of materials needed to complete it and easy-to-follow directions. Also, in many cases, examples are provided to even further simplify the directions. Gameboards and score sheets have been included where necessary, and an answer key is also provided in the back of the book.

You will be amazed at how much fun students can have completing these exciting activities as they master important math skills at the same time.

Name ____________________ Date ____________

everyday **problem solving**

TOP CARD

Materials: fifty-three 2 ½″ x 3 ½″ cards

Each card should contain four operation signs and two numbers. (See sample card below.) The two numbers shown on the top of the card will be in reverse order on the bottom of the card. On pages 3–5 are examples of cards you can make. Either enlarge the cards provided or make a new set of cards.

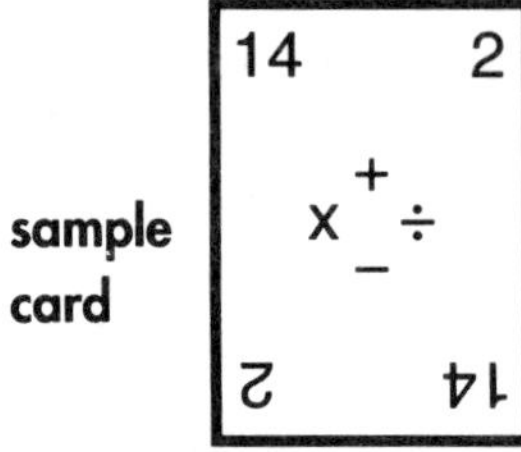

Larger (or smaller) numbers may be assigned to the cards depending on the level of difficulty you want.

Directions: Two players or two teams can play this game. Each team can have up to six players. The objective is to have the fewest number of cards left at the end of the game. The team with the least amount of cards at this point is declared the winner.

Shuffle the 53 cards and place them in a pile facedown. Turn the top three cards faceup and place each one in its own pile. The remaining 50 cards are then divided evenly among the two teams and dealt facedown. The teams keep the cards in a pile in front of them facedown.

Suppose that the top three cards are 6, 3; 14, 3; and 8, 2. They would appear as below:

A: 6 3 / + x ÷ − / 3 6

B: 14 3 / + x ÷ − / 3 14

C: 8 2 / + x ÷ − / 2 8

Name ______________________ Date ______________

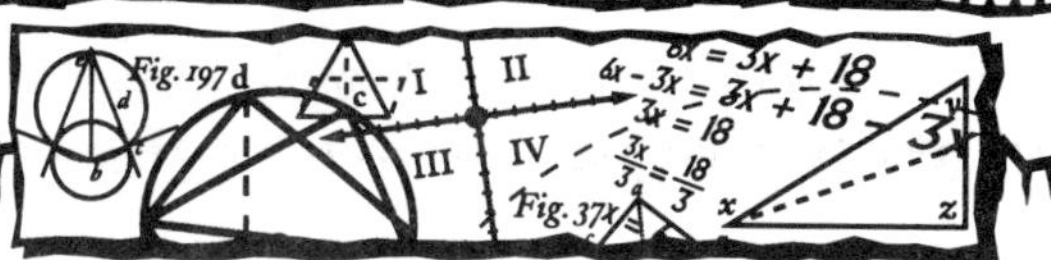

TOP CARD CONTINUED

Play then begins with each team taking a turn exposing its top card. Team One tries to match its card to any three of the cards faceup either using addition, subtraction, multiplication, or division. For example, suppose the top card exposed by Team One shows the numbers 4, 4. Team One may add, subtract, multiply, or divide 4 and 4, and it must call its match as well as the operation used. In this example, the match is at pile C, and the answer is 16. The explanation for this conclusion is 4 x 4 = 16 and 8 x 2 = 16. (The same operation does not have to be used for both cards.) Now, the new card 4, 4, is placed on top of card 8, 2. This card now controls pile C, and the other card(s) below it is forgotten.

It is now Team Two's turn. The cards on top of the piles are as follows:

A 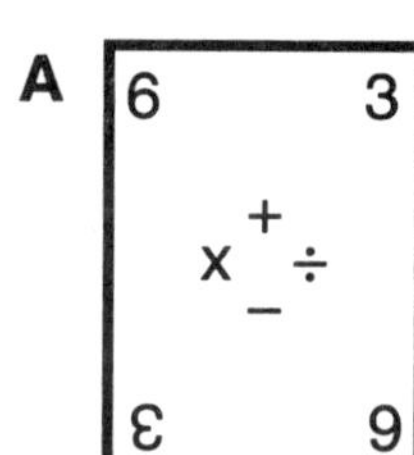

B 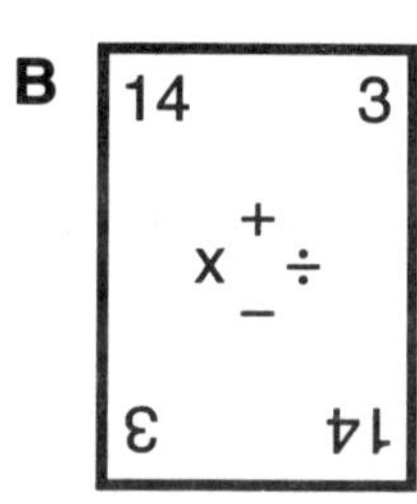

C 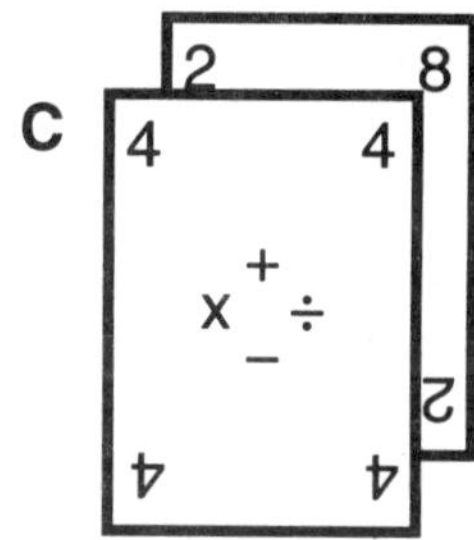

Team Two's top card is 4, 3. The matching pile is C, and the answer is 1. The explanation for this conclusion is 4 – 3 = 1 and 4 ÷ 4 = 1. Again, this new card 4, 3 is now placed on top of card 4, 4.

Play now returns to Team One. Team One turns over card 17, 2. After careful thought, this team must come to the conclusion that a match cannot be made with any of the three piles. Therefore, this card is to be placed to the right side of Team One.

As play continues for both teams, the top cards of the piles will change. It is the responsibility of each team to pay close attention to these changes. As the top card of each pile changes, it is up to team players to keep in mind the cards that have been previously placed to the right side. These cards may be ready to make a match and they may be placed without waiting for a team's turn. Only the cards on the side may be placed on a pile at any time. All remaining cards in the unexposed deck must await their turn.

Play continues until all the cards from the deck have been exposed. The winning team will have the fewest cards, and these are cards for which no match could be found.

Name ______________________________ Date ______________

everyday **problem solving**

Top Card Deck

5 4 + x ÷ − 4 5	4 3 + x ÷ − 3 4	8 7 + x ÷ − 7 8	6 6 + x ÷ − 6 6
5 5 + x ÷ − 5 5	4 4 + x ÷ − 4 4	3 3 + x ÷ − 3 3	3 2 + x ÷ − 2 3
10 10 + x ÷ − 10 10	9 9 + x ÷ − 9 9	8 8 + x ÷ − 8 8	11 2 + x ÷ − 2 11
11 1 + x ÷ − 1 11	11 3 + x ÷ − 3 11	14 3 + x ÷ − 3 14	10 6 + x ÷ − 6 10
	6 1 + x ÷ − 1 6	6 5 + x ÷ − 5 6	

Name ______________________________ Date ______________

everyday **problem solving**

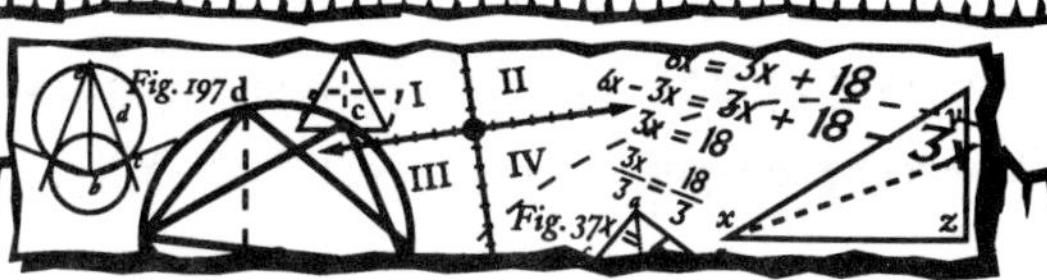

TOP CARD DECK

14 2 (x + – ÷)	14 1 (x + – ÷)	13 1 (x + – ÷)	15 2 (x + – ÷)
15 1 (x + – ÷)	16 4 (x + – ÷)	16 2 (x + – ÷)	16 1 (x + – ÷)
17 1 (x + – ÷)	18 3 (x + – ÷)	18 2 (x + – ÷)	18 1 (x + – ÷)
20 2 (x + – ÷)	20 3 (x + – ÷)	12 4 (x + – ÷)	12 3 (x + – ÷)
	12 2 (x + – ÷)	12 1 (x + – ÷)	

Name ______________________ Date ____________

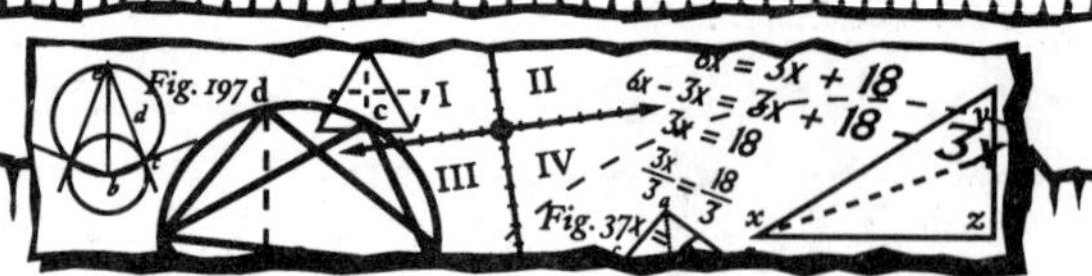

TOP CARD DECK

12 5	10 5	10 2	10 1
10 3	8 4	8 2	8 1
8 3	8 5	6 3	6 2
20 1	17 2	9 3	9 4
	9 2		

Name ______________________ Date ______________

THE PF GAMEBOARD

Materials: fifty-two 2 ½" x 3 ½" cards, 20 index cards

Make 52 playing cards. The chart shown tells how many of each card to make.

# on card	# of cards
2	14
3	12
5	12
7	7
11	7

On the index cards, make gameboards by writing one of the following numbers on each card: 30, 42, 44, 56, 70, 77, 84, 99, 100, 105, 110, 120, 125, 140, 150, 154, 200, 210, 231, and 330.

Sample Card and Gameboard:

Playing Card

2

3 ½"

2 ½"

Gameboard

30

index card

Directions: Two players or two teams can play this game. The objective is to have the largest score at the end of the game.

Place the gameboards in numerical order faceup on a table. Shuffle the 52 playing cards. Deal five cards to each team. The teams are to use these prime factor cards (playing cards) to create one of the products on the gameboards. When a team creates a product, the gameboard containing the product is removed from the table. The team is awarded that specific amount of points (gameboard).

The same number of prime factor cards (playing cards) that were required to make the selected product are given to the team. At all times, a team should have five playing cards placed in front of it for all participants to see. After a product has been selected and made, it then becomes the next team's turn.

Name ______________________ Date ______________

THE PF GAMEBOARD CONTINUED

Example: Team A's five playing cards are 11, 2, 5, 3, and 5. Team A chooses to go after gameboard 150. To accomplish this task, Team A would need to use the 2, 5, 3, and 5 from its exposed playing cards to create the equation below:

$$5 \times 5 \times 3 \times 2 = 150$$

Team A then earns 150 points and receives four more playing cards. Gameboard 150 is removed from the table. The playing cards and product cards that were used are then put in a discard pile.

Now it is Team B's turn. Team B's playing cards are 5, 7, 2, 3, and 2. Team B chooses to go after gameboard 210. Team B's equation is as follows:

$$5 \times 7 \times 2 \times 3 = 210$$

The equation is correct, and the 210 points belong to Team B. Since it used four of the playing cards, it will receive four new ones. The old playing cards are put in the discard pile.

The play is now under the control of Team A. The game continues until the remaining product cards can no longer be used to create an equation for a gameboard.

Name ______________________ Date ______________

everyday problem solving 

My Birthday Is the 156th Day

Materials: copies of page 9

Directions: This activity can be done individually or in small groups. The objective is to have students find a specific date in number terms.

Begin by telling students that January 1 is assigned the number 1, January 2 the number 2, and so on with February 1 being the number 32. (Make sure students use 365 days in a year.)

Have students think carefully about this important information. Eventually, they should begin to uncover an emerging pattern that assigns specific dates in the calendar year to a specific number.

Give each student a copy of page 9. Tell students to use their problem-solving skills to complete the page. Students are to assign a date (month and date) for each problem on the page. Remind students that January 1 is day 1.

Also included on the page are six challenge dates. You might want students to work in groups to solve these problems. As an extension, have pairs of students find six more challenge dates. Then pairs of students can exchange numbers and find the dates.

Name ______________________ Date ______________

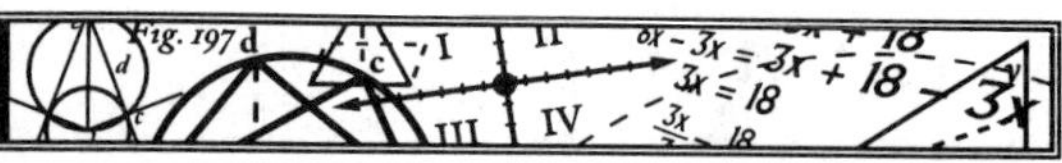

MY BIRTHDAY IS THE 156TH DAY

Find the month and day that is assigned to each number below.

a. 121 ______________

b. 59 ______________

c. 200 ______________

d. 156 ______________

e. 135 ______________

f. 222 ______________

g. 277 ______________

h. 340 ______________

i. 322 ______________

j. 257 ______________

k. 195 ______________

l. 100 ______________

m. 246 ______________

n. 47 ______________

o. 282 ______________

p. 300 ______________

q. 93 ______________

r. 139 ______________

s. 238 ______________

t. 95 ______________

Challenge Dates!

u. 553 ______________

v. 707 ______________

w. 392 ______________

x. 648 ______________

y. 421 ______________

z. 1005 ______________

Name ____________________ Date __________

USA, USA

Materials: fifty 2 ½″ x 3 ½″ cards, each containing one of the names of the 50 states on it; copies of "Probability Sheet and Awarded Points" (page 12)

The questions on page 12 reflect aspects of probability in regard to the 50 states. If the event of probability regarding a particular question is rather easy, then the number of points attributed to that question is small. If, on the other hand, the probability of a particular event occurring is more remote, then the number of points attributed to that question is much larger.

Alabama	**Louisiana**	**Ohio**
Alaska	**Maine**	**Oklahoma**
Arizona	**Maryland**	**Oregon**
Arkansas	**Massachusetts**	**Pennsylvania**
California	**Michigan**	**Rhode Island**
Colorado	**Minnesota**	**South Carolina**
Connecticut	**Mississippi**	**South Dakota**
Delaware	**Missouri**	**Tennessee**
Florida	**Montana**	**Texas**
Georgia	**Nebraska**	**Utah**
Hawaii	**Nevada**	**Vermont**
Idaho	**New Hampshire**	**Virginia**
Illinois	**New Jersey**	**Washington**
Indiana	**New Mexico**	**West Virginia**
Iowa	**New York**	**Wisconsin**
Kansas	**North Carolina**	**Wyoming**
Kentucky	**North Dakota**	

Name________________________________ Date____________

USA, USA CONTINUED

everyday problem solving

Directions: This game can be played with two to five teams. Each team can have up to five players. Each team will be given 100 points at the start of play. The 50 cards with the names of the 50 states are to be shuffled and placed facedown in a pile.

Before beginning the game, give each student a copy of page 12, "Probability Sheet and Awarded Points." Have students find the occurrence of the probability for each of the 15 events. Have students record their answers in both fractional and decimal notation. Students can use this sheet to help them play the game.

Teams take turns playing the game. The first player on Team One looks at the probability questions on page 12 and makes a determination as to what he or she thinks is on the top card in the pile. For example, suppose the player chooses question #3. He or she then picks the top card of the pile. If the card answers question #3, then Team One earns 12 points. This 12 points will be added to the original score of 100 giving Team One a total of 112 points. If the card doesn't answer the question, then Team One would have to subtract 12 points from its score giving the team a total of 88 points. Before the next team can take its turn, the entire deck of 50 cards must be reshuffled and the round continues as described.

The activity should allow for 15 rounds for each team. The team with the most points after the final round is the winner.

Extension: Another version of this game is to keep out the state cards already picked. This way, teams have to keep track of what cards have been picked and try to get a more accurate guess on the question to be answered.

Name______________________________ Date__________

PROBABILITY SHEET AND AWARDED POINTS

What is the probability of the top card in the deck being a state . . .

1. with the initial letter M?	6 points	**fraction** ________	**decimal** ________
2. with the initial letter N?	6 points	**fraction** ________	**decimal** ________
3. with the initial letter A?	12 points	**fraction** ________	**decimal** ________
4. with the initial letter C?	14 points	**fraction** ________	**decimal** ________
5. with the initial letter S?	15 points	**fraction** ________	**decimal** ________
6. with exactly ten letters in its name?	15 points	**fraction** ________	**decimal** ________
7. with exactly eight letters in its name?	3 points	**fraction** ________	**decimal** ________
8. with exactly four letters in its name?	14 points	**fraction** ________	**decimal** ________
9. with a vowel as the last letter in its name?	2 points	**fraction** ________	**decimal** ________
10. made up of two words?	3 points	**fraction** ________	**decimal** ________
11. that is one of the contiguous states?	1 point	**fraction** ________	**decimal** ________
12. with consecutive double letters?	6 points	**fraction** ________	**decimal** ________
13. whose total number of letters is a prime number of 5, 7, 11, or 13?	3 points	**fraction** ________	**decimal** ________
14. whose name has exactly three syllables?	2 points	**fraction** ________	**decimal** ________
15. that is your most favorite state?	40 points	**fraction** ________	**decimal** ________

everyday problem solving

Name________________________________ Date ______________

Mystic, Mindreader, Wizard, and Clairvoyant

Directions: This activity helps build students' self-esteem. Therefore, select one of your students who needs help in this area. Before the other students enter the class, give the student you chose the following information:

> *How would you like to be our mystic, mindreader, wizard, and clairvoyant? I will give you the same clue word each time a new number is selected by one of your classmates. You will not know what this number is, and you will not hear what it is. Yet, you will be able to tell your classmates the selected number. You will be able to do this because I will tell you the number in a secret way that your classmates will not know. Only you will be able to decode the selected number.*
>
> *I will let a classmate choose a number that is 26 or less. Once I know the number, I will let you know what it is without actually saying it. The trick to this is the important clue word that I will repeat each time I attempt to pass the number on to you. The clue word is READY. And it is the word immediately following READY that will require all of your attention. The initial letter of this follow-up word will tell you the number chosen by your classmate.*

As an example, explain to the student that the selected number is 7. You would say something like "Great mystic, mindreader, wizard, and clairvoyant Christopher, one does not question your powers Great One. Are you READY? Good, then tell everyone the selected number."

As the student picks up on the word READY and hears the next word which was *good,* then he or she will know that the initial letter in *good* is the letter g, and g is the seventh letter in the alphabet. Therefore, the secret chosen number is 7.

Name __________ Date __________

EVERYDAY *problem solving*

Mystic, Mindreader, Wizard, and Clairvoyant continued

As another example, you can say, "Great mystic, are you READY? Baseball is my favorite sport." The student hears the clue word READY and can easily figure out that the secret number is 2 because *baseball* begins with the second letter of the alphabet.

As the students become greatly impressed with the previously unknown abilities of their classmate, ask them to concentrate on just what it is you are doing. See if they can discover just how you are relaying the selected number to the mystic wonder. Inform them that you are doing the same thing every time a new secret number is chosen by the class.

This activity presents an excellent way to help students learn to solve a problem using trial and error.

Name ______________________ Date ______________

everyday problem solving

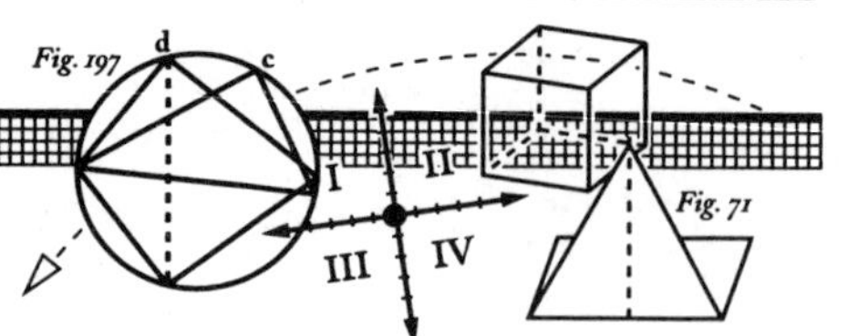

1001

Materials: eleven 2 ½″ x 3 ½″ cards, calculators (optional)

On each card, record the following:

1st card:	**Double the number.**
2nd card:	**Multiply the number by 5.**
3rd card:	**Subtract 16 from the number.**
4th card:	**Divide the number by 10.**
5th card:	**Add 14 to the number.**
6th card:	**Triple the number.**
7th card:	**Divide the number by 2.**
8th card:	**Multiply the number by 4.**
9th card:	**Multiply the answer at this point by 11.**
10th card:	**Multiply the answer at this point by 13.**
11th card:	**Multiply the answer at this point by 7.**

One to six players can play this game. Each player will get a turn working with each of the 11 cards.

Directions: Player One begins the activity by writing the number 26 on a sheet of paper. The first eight cards listed above are shuffled and placed facedown on the table. The top card is drawn and read aloud. Player One performs the operation as requested on the card. (Remember that the initial number is 26.) After an answer is reached, Player One will draw the second card.

Play continues in this manner until each of the first eight cards is drawn. (The order of these first eight cards does not matter.)

After Player One arrives at an answer after the eighth card is drawn, Player One is to round off his or her answer to the nearest three-digit number. For example, after the eighth card, an answer of 314.2 would be rounded off to 314, and an answer such as 67.48 would be rounded off to 67.5.

Name ________________________________ Date ______________

everyday problem solving

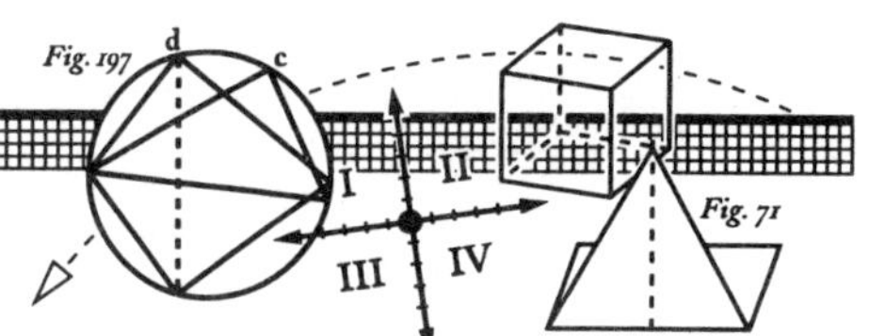

1001 CONTINUED

It is the last three cards (the ninth, tenth, and eleventh cards) that must be in the specific order as listed. It is also important that they be exposed at this juncture in the activity. Player One performs the operations on the ninth–eleventh cards.

After Player One has seen all 11 cards and arrived at an answer, all of the other players should write down Player One's answer. This same procedure is then to be followed by the remaining players. All players should begin with the number 26 and follow the same procedure.

Even though all of the players will start the activity with the same number, 26, it will be obvious that their answers will all probably be different. Remind all players to write each other's answers down. Students should see a pattern in the answers.

ADDITIONAL ACTIVITIES

- Have students use a calculator to multiply a three-digit number by 11, 13, and 7.
- Have students use a calculator to divide a six-digit number such as 714,714 by 1001.
- Have students use a calculator to multiply 515 by 10001, by 100001, and by 1000001.

Students should discover some very interesting patterns when completing these activities.

Name________________________ Date ____________

everyday **problem solving**

A Perfect Ten

Materials: nineteen 2 ½″ x 3 ½″ red cards, nineteen 2 ½″ x 3 ½″ black cards

Look at the chart below and make 38 cards containing the numbers 1, 2, 4, 8, and 10. Half of the cards should be red and half should be black. The red cards denote the numerators, and the black cards denote the denominators.

Number on Card

	1	2	4	8	10
red	6	4	4	3	2
black	6	4	4	3	2

Directions: This activity is for two players or two teams with up to five players on each team. The objective of the game is to reach a score as close to ten as possible without exceeding it.

Divide the 38 cards into red and black piles. Shuffle each pile and place each one facedown on a table. Players take turns picking the top card from each pile to make a fraction. As the fractions are made, they are added together. Each player will have up to ten opportunities to reach a perfect score of ten. Players will be given the option to stop play any time they feel they are close enough to ten points. If a player's score exceeds ten, then that player is automatically disqualified from further play.

For Example:

Player One picks the top red card and the top black card. Both cards are to be placed in the form of a fraction.

1st turn	red / black $\frac{4}{1}$	Player One has 4 points.
2nd turn	red / black $\frac{1}{8}$	The new fraction is now added to the previous one. Therefore, at the end of the second turn, Player One has 4 ⅛.
3rd turn	red / black $\frac{2}{4}$	After this turn, Player One has 4 ⅝.
4th turn	red / black $\frac{4}{1}$	Player One now has 8 ⅝.

Name ______________________ Date ______________

A PERFECT TEN CONTINUED

Player One goes into the 5th turn with a score of 8 5/8.

Player One thinks he or she is close enough to The Perfect Ten and wishes to stop play at this time and declare this score.

Player One returns all of the red and black cards to the original piles. Each pile is shuffled and placed facedown on the table for Player Two's turn. Player Two is now aware of the score that he or she must exceed in order to win the game (8 5/8), keeping in mind that The Perfect Ten must not be surpassed.

Extension: After students play the game a number of times, you may wish to change the card distribution to the following:

Number on Card

	1	2	3	4	5	8
red	6	4	4	2	2	1
black	4	2	2	2	2	1

The objective remains the same, and that is the goal of The Perfect Ten.

Name________________________________ Date ______________

Decode, Decipher, and Translate to an Understanding

Materials: nineteen 3″ x 3″ cards

Make the 19 cards following the patterns on page 21. Make sure you put a letter at the top of each card. Additional cards can be made to make the game more challenging.

Directions: Two to five players, or two teams with up to five players on each team, can play this game. Shuffle the cards and place them facedown in a pile. Once all of the players are ready, one person takes the top five cards and spreads them out faceup on a table for all the players to see. Once the five cards are placed on the table, they may not be touched or moved by any player. Touching or moving a card will break the concentration of other players who may be considering patterns.

The objective of the game is for all players to examine the five cards on the table and look for a pattern for each card. Players may look at the numbers in any way they wish. The three numbers on each card may be looked at in terms of place value or as 12 single-digit numbers that happen to be on the card. The order of the numbers may not be changed. There may be more than one pattern associated with each card, but the numbers must show some type of relationship.

When a player is ready to claim a card, he or she must identify the card by the letter in the upper left-hand corner. Once the card has been identified, the player relates the pattern that he or she has discovered. Once all the other players agree with the discovery, then that card and the ensuing point(s) associated with it will belong to the player. If the pattern matches the one recorded in the answer key, then the player is awarded one point. If the player has discovered a different but also correct answer regarding the numbers on the card, then the player will be awarded two points.

Name ______________________ Date ____________

DECODE, DECIPHER, AND TRANSLATE TO AN UNDERSTANDING CONTINUED

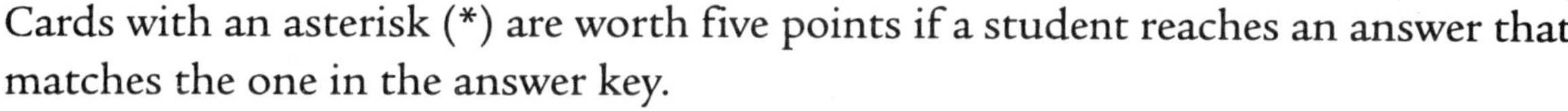

Cards with an asterisk (*) are worth five points if a student reaches an answer that matches the one in the answer key.

After points are given, the card is removed from the table, and a new card from the top of the deck is put in its place. Play continues until all 19 cards have been correctly given patterns.

Tell students to be ready to examine each card with "open eyes." They should approach this pattern search with a creative mind as there are many correct answers to the problems presented.

Example of a Turn

A player calls out card z. He or she says card z has three four-digit numbers that are greater than 9,000 but less than 10,000. All the other players agree with this discovery. The answer key is checked, and the same pattern is revealed. Therefore, this player gets one point, and the card is removed. A new card from the facedown pile is picked and put in its place and play resumes.

z

9 1 3 8

9 2 4 6

9 7 7 1

Name ______________________________ Date ______________

DECODE, DECIPHER, AND TRANSLATE CARDS

Card	Row 1	Row 2	Row 3
a	3 2 4 7	4 5 9 8	2 7 7 6
b	8 8 4 4	8 4 4 8	4 4 8 8
c	7 1 3 8	5 9 1 2	1 7 9 4
d*	1 7 7 6	1 8 1 2	1 4 9 2
e*	6 2 3 1	8 8 4 4	4 6 2 3
f	5 6 7 8	2 3 4 5	6 7 8 9
g	4 0 1 3	8 0 7 9	6 0 4 8
h	9 9 6 8	6 9 9 8	6 8 9 9
i	9 7 7 3	3 5 4 4	6 6 7 3
j	4 7 7 4	3 1 1 3	9 4 4 9
k	2 5 6 4	3 6 4 9	8 1 1 6
l	5 1 2 5	7 3 4 7	6 2 1 6
m*	9 6 3 6	1 5 1 5	3 6 1 2
n*	8 5 2 1	6 3 0 3	7 1 1 5
o*	8 1 8 3	4 3 4 5	2 1 8 1
p*	4 0 1 4	9 0 0 0	3 1 3 2
q	3 9 9 6	9 3 3 3	6 3 6 6
r*	1 1 2 3	5 8 1 3	2 1 3 4
s*	1 3 5 7	2 0 4 0	1 6 0 3

Name ______________________ Date ______________

everyday problem solving

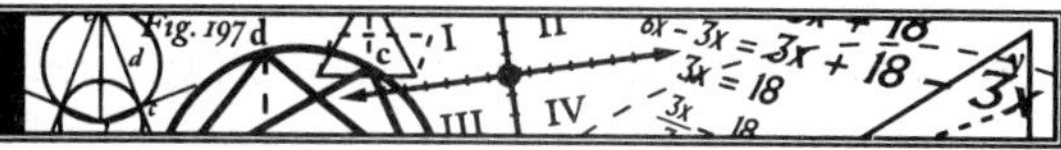

WINNER TAKES ALL!

Materials: fourteen 5″ x 8″ cards, aces through tens in a deck of playing cards (40 cards total)

Make 14 answer cards (5″ x 8″) by writing one of the following on each card:

- Largest factor of 40
- Largest multiple of 5
- Largest odd number
- Smallest prime number
- Largest answer in the 30s
- Largest multiple of 12
- Smallest composite number in the 20s
- Answer closest to 0
- Largest even number
- Largest prime number
- Smallest even number
- Largest factor of 50
- Largest odd number in the 60s
- Largest answer closest to 90

Directions: This game is for six players or six teams with each team consisting of two to three players. The 40 playing cards should be shuffled. Each team should be dealt three playing cards.

The 14 answer cards are shuffled and placed facedown in the middle of the playing area. When play is ready to begin, the top answer card is flipped over for the teams to see.

Each team must use all three of its playing cards and arrive at an equation that solves the answer card. The team may use one operation or combine operations. The team members may only use each card one time. The team closest to the answer wins all of the cards from each team. (Each card counts as one point.) In case of a tie, the cards should be shared evenly among the teams that had the same answer.

Name________________________________ Date ______________

everyday problem solving 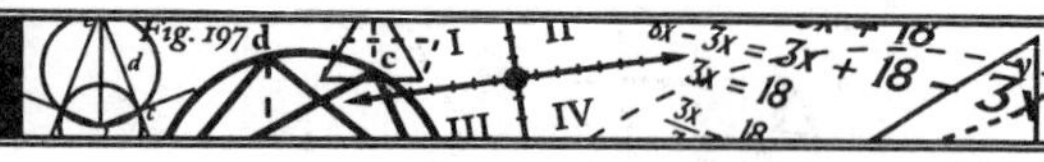

WINNER TAKES ALL! CONTINUED

Play continues in this manner. Three new cards are dealt to each team, and a new answer card is flipped over.

Example: Below are six teams and their playing cards.

A	B	C	D	E	F
5 9 1	6 2 3	6 5 7	3 4 3	3 8 2	5 2 7

The answer card says "largest even number." The equations the six teams might come up with are as follows:

Team A	9 x (5 + 1) = 54
Team B	(6 x 3) x 2 = 36
Team C	7 x (5 x 6) = 210
Team D	3 x 3 x 4 = 36
Team E	8 x (3 x 2) = 48
Team F	(5 x 2) x 7 = 70

Team C would be the winner of this round and collect all of the 18 playing cards for a total of 18 points.

Name________________________________ Date____________

STATE YOUR ANSWER

Materials: sixty $2\frac{1}{2}'' \times 3\frac{1}{2}''$ cards

On each of the 60 cards, the following should be written:

15 addition facts written with a blue marker

15 subtraction facts written with a red marker

15 multiplication facts written with a green marker

15 division facts written with a black marker

All of the cards should reflect the basic math facts needed to be reinforced by your students. All 60 cards should be arranged in the format shown to the left. The arrangement on the face of each card will allow all of the players equal opportunities to recognize and acknowledge the basic fact that is being presented.

Directions: This game is for six players. Each of the six players is given a state name of Maine, Maryland, Michigan, Minnesota, Mississippi, Missouri, or Montana.

Shuffle the 60 cards. Deal 10 cards to each player. All of the cards should be kept in a pile facedown. The objective of the game is to have the least amount of cards at the end of the game.

To begin the game, each player must turn over his or her top card. The players race to try to match their top card with another player's top card (same answers are matches). This match can happen at any time. When a player sees a match, he or she must call the match correctly.

everyday problem solving

Name________________________ Date__________

STATE YOUR ANSWER CONTINUED

everyday problem solving

To call a match correctly, the player must first identify the name of the match's state and the correct answer of the match. If either is incorrect, then the player with the error must accept all of the exposed cards from the other player. If both are correct, then the player gives his or her cards to the opposing player. (A player might have more than one exposed card. A player could have several if he or she hasn't been able to match answers with another player. If this happens, when the player does have a match and it is correct, the player gives all of his or her exposed cards away.)

Example: Perhaps a match is spotted between the Michigan player and the Maryland player, and the matching answer on the top of both exposed piles is 36 (12 x 3 for one player and 6 x 6 for the other player). Spotting this match, the Michigan player must say, "Maryland 36" before the Maryland player says, "Michigan 36." If the Michigan team did this, then it would now give its pile of exposed cards to the Maryland player. At this point, it might be best if you were to act as a referee. This situation should continue until the game becomes more familiar to the players.

Name ____________________ Date ____________

Make It, Then Take It

Materials: forty $2\frac{1}{2}'' \times 3\frac{1}{2}''$ cards, ten $5'' \times 8''$ index cards

On each of the 40 cards, write the numbers 1–10 (four cards with each number). On each index card, write one of the following numbers: 29, 39, 83, 47, 38, 58, 70, 67, 75, and 90.

Directions: This game is played with two teams. Place the 10 index cards faceup on a table so that each team is able to see all 10. Shuffle the 40 cards and deal each team four cards faceup.

The object of the game is for each team to use all four of its cards (numbers) to create an equation with the answer being on one of the index cards. If a team is able to create an equation with an answer that is the same as on one of the index cards, then this team will earn that many points.

Five rounds of play will constitute the game. Each round will consist of anywhere from one to two minutes. After this time period has elapsed and no team has created an equation, then the exposed playing cards from each team will be collected, and the entire deck will be reshuffled. The next round continues with four new cards for each team.

Example round:

Team A's cards are 8, 4, 5, and 10. The equation it came up with is $(8 \times 4) + (10 + 5) = 47$. Team A will earn 47 points because the equation is correct, all four numbers were used, and the result of the equation matches one of the index cards.

Team B's cards are 7, 9, 8, and 4. The equation it came up with is $(7 \times 9) + (8 - 4) = 67$. Team B has earned 67 points.

This is the end of the first round. All cards are then shuffled, and each team will receive four new cards, a fresh time limit, and another chance to form winning equations. At the end of the fifth round, the team with the greatest combined point score will be declared the game winner.

Name________________________________ Date ______________

everyday

problem solving

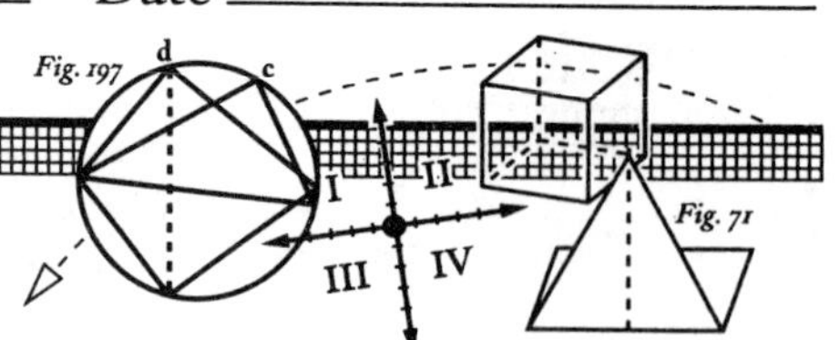

Closest to 101

Materials: thirty-six 2½" x 3 ½" cards

On each card, write the numbers 1–9.
(Each number should be written on four cards.)

This game is for two to five players or two to five teams. Each player is given 101 points to start the game. The objective of the game is to have a score as close to 101 as possible.

To begin, the 36 cards are shuffled and given to Player One in a stack facedown. On a sheet of paper, Player One writes down the initial number of 101. The one rule of this game is that each player must use addition three times and subtraction four times. The operations may be used in any order. Player One turns over the top two cards from the deck. Suppose the two cards are 9 and 3. Player One decides if he or she wants to use the number 93 or 39 (a two-digit number is always used) and which operation he or she wants to use. Player One decides to add 39. Therefore, the first step is the equation 101 + 39 = 140. Continuing in the same manner, the six steps of Player One are as follows:

Step 1	3, 9	101 + 39 = 140
Step 2	5, 3	140 – 35 = 105
Step 3	1, 5	105 – 15 = 90
Step 4	7, 8	90 + 78 = 168
Step 5	7, 2	168 – 72 = 96
Step 6	5, 4	96 + 45 = 141

Player One has used addition and subtraction each three times. Therefore, the last step must be the final subtraction as well as the last opportunity to come as close as possible to the score of 101.

Step 7	6, 3	141 – 36 = 105

Name ______________________ Date ______________

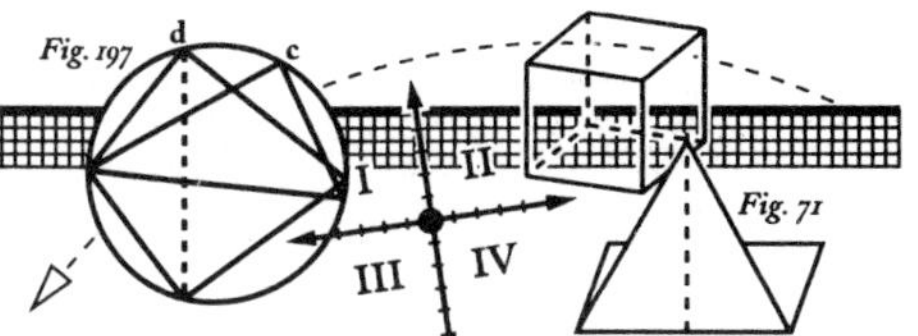

Closest to 101 continued

Player One's turn is over, and his or her final score is 105. The other players realize that they must come closer to 101 than 105 to win.

The 36 cards are reshuffled, and it is now Player Two's turn. After all the players have had a turn, the winner is the player whose score is closest to 101.

Extensions: Use three cards and an initial target number of 1001, or four cards and an initial target number of 10,001.

Another possibility with a target number of 101 would be to include one multiplication operation in addition to the three addition and four subtraction operations. The objective would still be the same and that is the target number of 101.

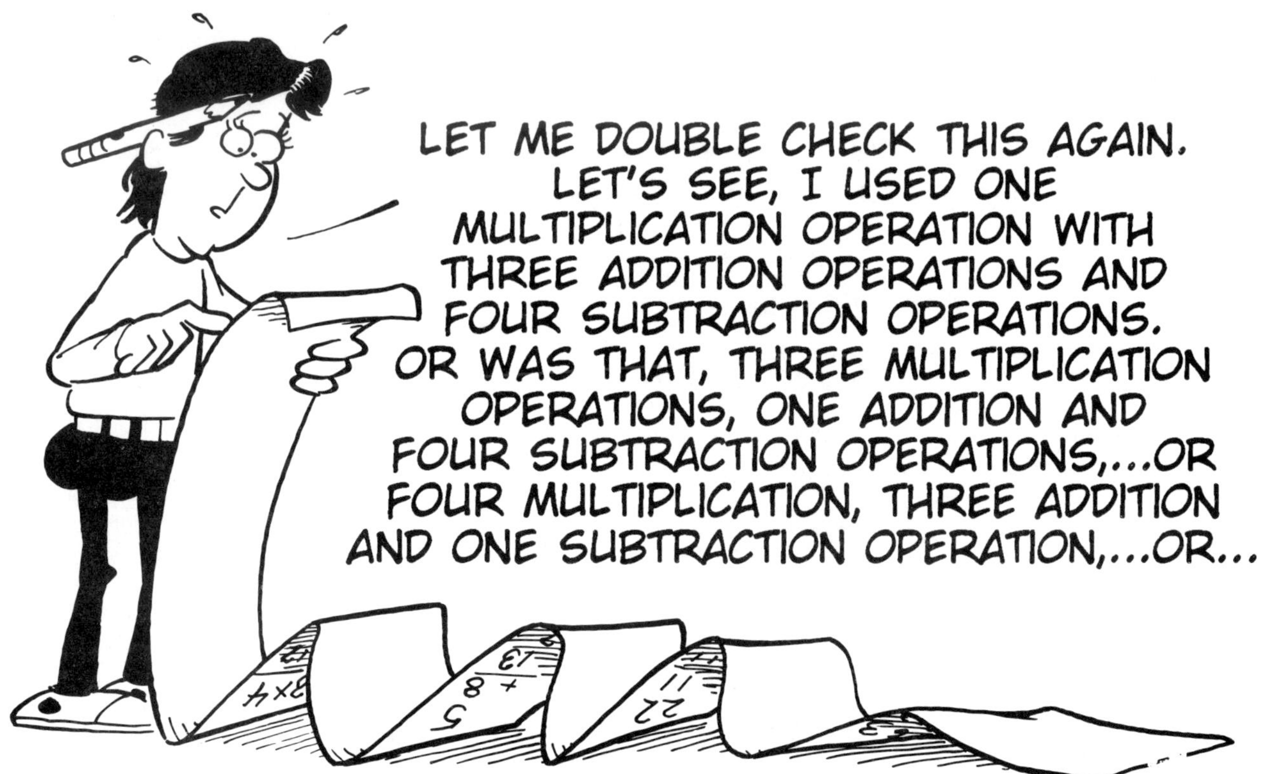

Name ______________________ Date ______________

everyday **problem solving**

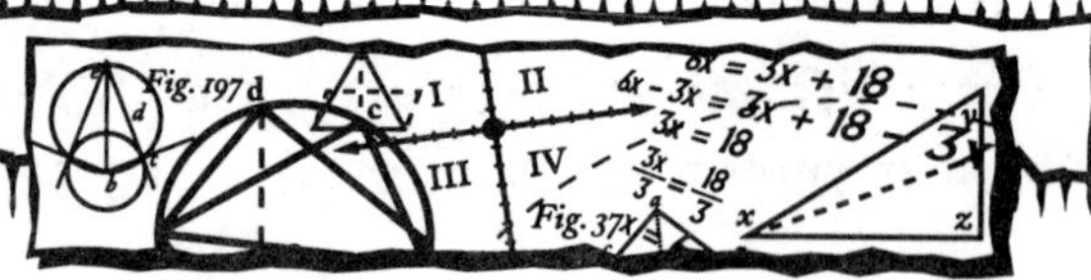

The Ice Cream Sundae Business

Materials: copies of page 31

Directions: Go over the situation below with the students before giving them copies of page 31 to complete. Make sure students understand the reasoning on how each order was filled. If some students are having difficulty with the assignment, let students work in cooperative learning groups.

The new owner of an ice cream parlor came upon a creative way to increase the volume of her business and hopefully end up with a decent profit. To increase proficiency in delivering her product (ice cream sundaes), she developed and implemented the method described below:

> When a customer orders his or her choice of an ice cream sundae with the choice of toppings, he or she refers to the chart listed below. The customer can see a chart containing seven columns. The columns are listed in this specific order with the choice of ice cream as the first selection and sprinkles listed as the last choice.

1	2	4	8	16	32	64
1 x 1	2 x 1	2 x 2	2 x 2 x 2	2 x 2 x 2 x 2	2 x 2 x 2 x 2 x 2	2 x 2 x 2 x 2 x 2 x 2
ice cream	hot fudge	caramel	nuts	bananas	marshmallows	sprinkles

(Note the prime factors and the products above each choice of topping. The numeral 1 above the ice cream is neither prime nor composite.)

Suppose a customer chooses nuts, hot fudge, and ice cream. Instead of wasting a great deal of precious time, the customer would say to the employee behind the counter, "I would like 11, please!"

Name ______________________ Date ______________

everyday **problem solving**

THE ICE CREAM SUNDAE BUSINESS CONTINUED

With the system created by the shop owner, the employee hears and then locks in on the number 11. He or she then accordingly knows that the only way to create 11 is to think of its components, namely, the numbers 8, 2, and 1. And of course due to an intense training program, the employee knows to equate the 8 with nuts, the 2 with hot fudge, and 1 with ice cream. Suppose another customer came in and ordered a 23. The sundae server knows that 23 is comprised of these components 16, 4, 2, and 1. The server then knows the sundae to be served will contain bananas (16), caramel (4), hot fudge (2), and ice cream (1). Once the sundae is given to the customer, the employee should always remember to say, "Please enjoy!"

Extensions: Have students work in cooperative learning groups and pretend they all decided to go into a business. They should also pretend they want to develop a format similar to the one successfully used in the Ice Cream Sundae Business. Have the groups open up a pizza place.

The groups should set up the following number ordering format: 64, 32, 16, 8, 4, 2, 1. Remind them to list the toppings below each number. Have each group try the system out on the rest of the class.

Name ____________________ Date ____________

everyday **problem solving**

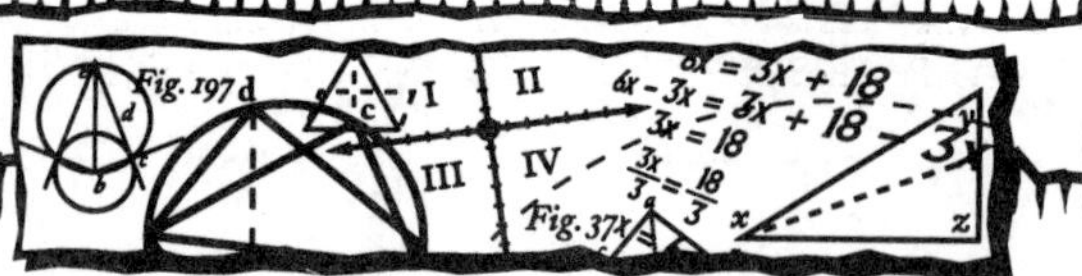

THE ICE CREAM SUNDAE BUSINESS

The problems listed below are intended to help you improve your problem solving and decoding skill abilities. Pretend you are the counter person at the Ice Cream Sundae Parlor and you are in a very busy situation. Fill out the orders below. Remember:

1	2	4	8	16	32	64
1 x 1	2 x 1	2 x 2	2 x 2 x 2	2 x 2 x 2 x 2	2 x 2 x 2 x 2 x 2	2 x 2 x 2 x 2 x 2 x 2
ice cream	hot fudge	caramel	nuts	bananas	marshmallows	sprinkles

a. 19 ____________________

b. 35 ____________________

c. 21 ____________________

d. 37 ____________________

e. 25 ____________________

f. 13 ____________________

g. 27 ____________________

h. 41 ____________________

i. 49 ____________________

j. 23 ____________________

k. 65 ____________________

l. 51 ____________________

m. 15 ____________________

n. 31 ____________________

o. 97 ____________________

p. 33 ____________________

q. 29 ____________________

r. 101 ____________________

At one point during a very busy summer evening in the ice cream parlor, five customers came in to order their own special sundaes. They almost ruined the morale of the store owner when they gave her the sundae orders of 28, 22, 40, 34, and 48. Why was the owner so upset with these five orders?

Name______________________________ Date ______________

WHAT?

Materials: copies of page 33

Directions: This activity has been designed to help strengthen student development and appreciation of problem-solving techniques. It is important for students to learn how to listen, think, and then be able to solve problems.

When most students are asked to solve the problem 5 + 3, the answer is almost always 8.

For this activity, however, the question is 5 + 3 = 2 what? One situation where the answer to this question could be 2 is when we think about money—
5 quarters + 3 quarters = 2 dollars.

Give each student a copy of page 33. Students can either work individually or in small groups to complete the activity.

Name ______________________________ Date ______________

EVERYDAY PROBLEM SOLVING

WHAT?

Solve the problems below.

Example: 5 + 3 = 2 what?

5 quarters + 3 quarters = 2 dollars

A. 17 + 4 = 3 what?

____

B. 27 + 9 = 1 what?

____

C. 40 + 8 = 3 what?

____

D. 725 + 5 = 2 what?

____

E. 13 + 11 = 2 what?

____

F. 119 + 25 = 1 what?

____

G. 1100 + 900 = 1 what?

____

H. 86 + 14 = 1 what?

____

I. 15 + 5 = 2 what?

____

J. 33 + 15 = 4 what?

____

Name ____ Date ________________

everyday problem solving

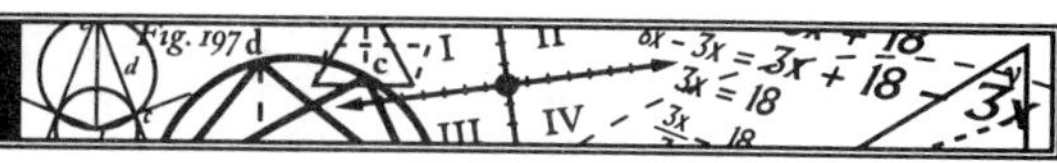

IFOYE (IN FRONT OF YOUR EYES)

Materials: copies of pages 35–36

Directions: On the board, write examples **a** and **b** below. Have students examine the examples to see if they can find the hidden math word that is in each sentence. Tell students that the letters making up the word are in order, but the letters may be separated by one or more spaces. If students are having difficulty finding the words, you could give them hints. For example, in example **a** below, a clue might be "difference."

Make copies of pages 35–36 for each student. Have students work individually or in pairs to find the hidden math words in the sentences. Students will need to use careful observation skills when working on this activity. As you circulate to see if students are having problems, give helpful hints. Or, pairs of students can give hints to each other.

Examples:

a. The gem in using good study habits is the reward of better grades.

b. If actor Tom Hanks wins the award, attendance at this movie will drastically increase.

In example **a**, the math word is *minus.* The ge**m in us**ing . . .

In example **b**, the math word is *factor.* **If actor** Tom . . .

Name ________________________________ Date ______________

everyday problem solving

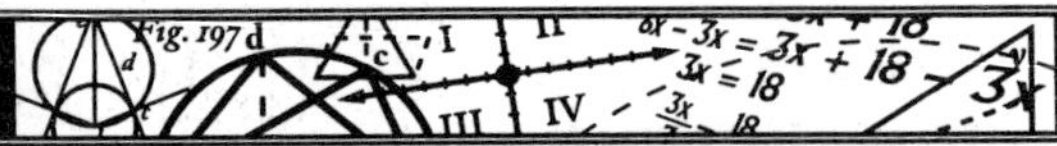

IFOYE (In Front of Your Eyes)

After reading the sentences below and on page 36 very carefully, you should be able to uncover a secret, well-concealed math word. Each sentence contains its own mystery word. The spelling of each concealed math word is contrived so that the letters required to spell the word are intact. The letters in the secret word are, however, separated only by one or more spaces. This means that if you were to remove the space between the letters, you would immediately recognize the math word.

Look for the hidden math word in each sentence below and on page 36 and record it on the corresponding line.

1. Their house venture turned out to be most interesting. ______________
2. The beautiful young cat would purr every evening. ______________
3. Mia said, "It was hers I x-rated before giving the test." ______________
4. It is all of us who need math in our lives. ______________
5. What entertainment will keep you awake? ______________
6. It is Sir Artwel Vernon who returned to England last year. ______________
7. If I've never been there, then how would I know who she is? ______________
8. It was Robyn who said to her brother Alan, "Can I net your lobster?" ______________
9. Both reeds are in the large symphony orchestra. ______________
10. You can be sure that if our team wins, we'll throw a party. ______________
11. The ship was required to shun dredging that was being conducted in the middle of the river.

    ____
12. The hotel "Evenings End" looked great to Larry after the long journey.

    ____

Name ______________________ Date ______________

everyday problem solving 

IFOYE (In Front of Your Eyes) CONTINUED

13. It seems that Mark and Scott went yearly after the initial experience. ______________

14. "Neigh to all this," whined the magnificent horse. ______________

15. The book was really great. It truly was, even Tyrone agreed. ______________

16. Jeff, if Tyrell approves, will you go to the concert with us? ______________

17. "Can I never be by myself?" asked Leonora. ______________

18. The freight you asked about was shipped last Monday. ______________

19. The treat is for one of their dogs, Keesha, Puff, or Tyke. ______________

Name____ Date______________

THE BAR CODE

everyday problem solving

Materials: copies of page 38

Directions: The bar code is a computational system that is derived from the base three system. The following are bar code numbers and their base ten counterparts:

= 1	= 7
= 2	= 8
= 3	= 9
= 4	= 10
= 5	= 27
= 6	= 81

The bar code system is based on the operation of addition as well as the concept of place value. Its most important bar code markings are 1, 9, 27, and 81.

On the board, write the above bar codes with their numerical values. Go over each bar code with the students. Explain to students how each number is derived. Then give each student a copy of page 38. Have students work individually or in groups to complete the page.

As an extension, have students make up their own bar codes. Then they can trade bar codes with classmates and solve each other's.

Name ______________________________ Date ____________

THE BAR CODE

Write the numerical base ten value for each bar code below.

a. = ____

b. = ____

c. = ____

d. = ____

e. = ____

f. = ____

g. = ____

h. = ____

i. = ____

j. = ____

k. = ____

l. = ____

m. = ____

n. = ____

o. = ____

p. = ____

q. = ____

r. = ____

s. = ____

t. = ____

everyday problem solving

Name ______________________________ Date ______________

THE SUPERSTARS

Materials: copies of pages 40–44

Directions: On page 41 is a chart comprised of the names of 63 professional athletes from the sports of baseball, basketball, football, and hockey. The names of some of the students' favorite players may be among those listed. This activity will allow students to learn the names of each other's favorite athletes just by asking a few simple questions that relate to the names on the chart. The procedure, in which students will be able to tell someone the name of his or her star player, is not magical. The method of describing how to find the name is itself an important problem-solving skill.

The names of the 63 players are listed in columns A–F (pages 42–44). Most players are listed in more than one column. The columns are given the following numerical values: A = 32, B = 16, C = 8, D = 4, E = 2, and F = 1. These numbers are very important in completing this activity.

Instruct a student to select any one of the 63 athletes on the chart on page 41. Tell the student not to tell you the name he or she selected. Instead, have the student look at the columns A–F and tell you the column(s) his or her player is listed in. Suppose the student tells you his or her favorite player can be found in columns B, E, and F. You mentally note that B = 16, E = 2, and F = 1. Adding these numbers together, you get a total of 19. If you look on page 41, you discover that 19 equals Julius Erving which should be the athlete the student chose.

Some more examples include these:

A, B, D (32 + 16 + 4 = 52) = Pete Rose

C, D, E, F (8 + 4 + 2 + 1 = 15) = Ty Cobb

A, C, F (32 + 8 + 1 = 41) = Joe Namath

Give students copies of page 40. Have students complete the page individually. Once students are "superstars" at this activity, give them copies of pages 41–44 and have them try this activity with their friends and family members.

Name ______________________ Date ______________

THE SUPERSTARS

Find the players listed below.

1. B ______________
2. A, C, F ______________
3. C, D, F ______________
4. B, F ______________
5. D, E, F ______________
6. A, D, F ______________
7. A, B, C ______________
8. E, F ______________
9. B, C, D ______________
10. A, C, E, F ______________
11. B, C, E ______________
12. A, E, F ______________
13. D, E ______________
14. C, E, F ______________
15. A, B, C, D ______________
16. A, D, E ______________
17. B, C, D, F ______________
18. A, B, C, D, E ______________
19. C, F ______________
20. D, F ______________
21. B, D, E ______________
22. C, D, E ______________
23. A, B, C, D, F ______________
24. A, D, E, F ______________
25. Cal Ripkin, Jr.'s name is under which column(s)? ______________
26. Cecil Fielder's name is under which column(s)? ______________
27. Nolan Ryan's name is under which column(s)? ______________
28. Bobby Orr's name is under which column(s)? ______________

Name ______________________ Date ______________

EVERYDAY *problem solving*

THE SUPERSTARS CHART

1. Hank Aaron
2. Troy Aikman
3. Ernie Banks
4. Charles Barkley
5. Larry Bird
6. Drew Bledsoe
7. Wade Boggs
8. George Brett
9. Jim Brown
10. Jose Canseco
11. Steve Carlton
12. Wilt Chamberlain
13. Bobby Clarke
14. Roberto Clemente
15. Ty Cobb
16. Randall Cunningham
17. Joe DiMaggio
18. Clyde Drexler
19. Julius Erving
20. Patrick Ewing
21. Cecil Fielder
22. Jimmy Foxx
23. Lou Gehrig
24. Wayne Gretzky
25. Kareem Abdul-Jabbar
26. Magic Johnson
27. Michael Jordan
28. Sandy Koufax
29. Mario Lemieux
30. Eric Lindros
31. Satchel Paige
32. Kirby Puckett
33. Karl Malone
34. Mickey Mantle
35. Dan Marino
36. Roger Maris
37. Willie Mays
38. Joe Montana
39. Warren Moon
40. Stan Musial
41. Joe Namath
42. Hakeem Olajuwon
43. Shaquille O'Neal
44. Bobby Orr
45. Walter Payton
46. Scottie Pippen
47. Jerry Rice
48. Cal Ripkin, Jr.
49. Brooks Robinson
50. David Robinson
51. Jackie Robinson
52. Pete Rose
53. Nolan Ryan
54. Ryne Sandberg
55. Mike Schmidt
56. Emmitt Smith
57. Fran Tarkenton
58. Lawrence Taylor
59. Johnny Unitas
60. Reggie White
61. Dominique Wilkins
62. Ted Williams
63. Steve Young

Name ____________________ Date ____________

EVERYDAY *problem solving*

THE SUPERSTARS CHART CONTINUED

Karl Malone
Mickey Mantel
Dan Marino
Roger Maris
Willie Mays
Joe Montana
Warren Moon
Stan Musial
Joe Namath
Hakeem Olajuwon
Shaquille O'Neal
Bobby Orr
Walter Payton
Scottie Pippen
Kirby Puckett
Jerry Rice
Cal Ripkin, Jr.
Brooks Robinson
David Robinson
Jackie Robinson
Pete Rose
Nolan Ryan
Ryne Sandberg
Mike Schmidt
Emmitt Smith
Fran Tarkenton
Lawrence Taylor
Johnny Unitas
Reggie White
Dominique Wilkins
Ted Williams
Steve Young

Kareem Abdul-Jabbar
Randall Cunningham
Joe DiMaggio
Clyde Drexler
Julius Erving
Patrick Ewing
Cecil Fielder
Jimmy Foxx
Lou Gehrig
Wayne Gretzky
Magic Johnson
Michael Jordan
Sandy Koufax
Mario Lemieux
Eric Lindros
Satchel Paige
Cal Ripkin, Jr.
Brooks Robinson
David Robinson
Jackie Robinson
Pete Rose
Nolan Ryan
Ryne Sandberg
Mike Schmidt
Emmitt Smith
Fran Tarkenton
Lawrence Taylor
Johnny Unitas
Reggie White
Dominique Wilkins
Ted Williams
Steve Young

Name ____________________ Date ____________

EVERYDAY *problem solving*

THE SUPERSTARS CHART CONTINUED

Kareem Abdul-Jabbar
George Brett
Jim Brown
Jose Canseco
Steve Carlton
Wilt Chamberlain
Bobby Clarke
Roberto Clemente
Ty Cobb
Wayne Gretzky
Magic Johnson
Michael Jordan
Sandy Koufax
Mario Lemieux
Eric Lindros
Stan Musial
Joe Namath
Hakeem Olajuwon
Shaquille O'Neal
Bobby Orr
Satchel Paige
Walter Payton
Scottie Pippen
Jerry Rice
Emmitt Smith
Fran Tarkenton
Lawrence Taylor
Johnny Unitas
Reggie White
Dominique Wilkins
Ted Williams
Steve Young

Charles Barkley
Larry Bird
Drew Bledsoe
Wade Boggs
Wilt Chamberlain
Bobby Clarke
Roberto Clemente
Ty Cobb
Patrick Ewing
Cecil Fielder
Jimmy Foxx
Lou Gehrig
Sandy Koufax
Mario Lemieux
Eric Lindros
Roger Maris
Willie Mays
Joe Montana
Warren Moon
Bobby Orr
Satchel Paige
Walter Payton
Scottie Pippen
Jerry Rice
Pete Rose
Nolan Ryan
Ryne Sandberg
Mike Schmidt
Reggie White
Dominique Wilkins
Ted Williams
Steve Young

Name ____________________ Date ____________________

EVERYDAY *problem solving*

The Superstars Chart continued

Troy Aikman
Ernie Banks
Drew Bledsoe
Wade Boggs
Jose Canseco
Steve Carlton
Roberto Clemente
Ty Cobb
Clyde Drexler
Julius Erving
Jimmy Foxx
Lou Gehrig
Magic Johnson
Michael Jordan
Eric Lindros
Mickey Mantle
Dan Marino
Joe Montana
Warren Moon
Hakeem Olajuwon
Shaquille O'Neal
Satchel Paige
Scottie Pippen
Jerry Rice
David Robinson
Jackie Robinson
Ryne Sandberg
Mike Schmidt
Lawrence Taylor
Johnny Unitas
Ted Williams
Steve Young

Hank Aaron
Kareem Abdul-Jabbar
Ernie Banks
Larry Bird
Wade Boggs
Jim Brown
Steve Carlton
Bobby Clarke
Ty Cobb
Joe DiMaggio
Julius Erving
Cecil Fielder
Lou Gehrig
Michael Jordan
Mario Lemieux
Karl Malone
Dan Marino
Willie Mays
Warren Moon
Joe Namath
Shaquille O'Neal
Satchel Paige
Walter Payton
Jerry Rice
Brooks Robinson
Jackie Robinson
Nolan Ryan
Mike Schmidt
Fran Tarkenton
Johnny Unitas
Dominique Wilkins
Steve Young

Name________________________________ Date ______________

everyday

problem solving

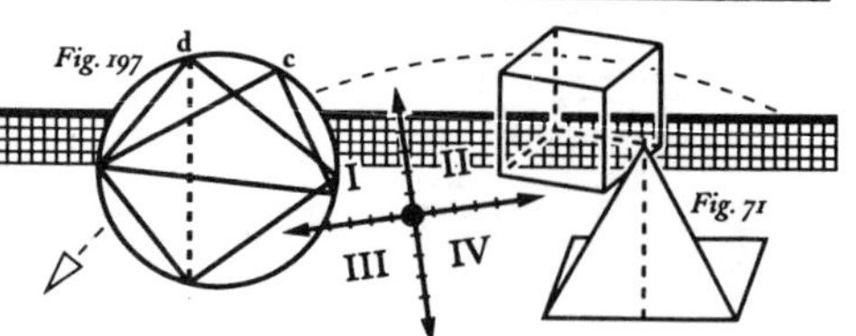

FRAC-CODE

Materials: copies of pages 47 and 48

Directions: On the board, write the following portion of the alphabet with the corresponding point values. This is to serve as a reference point for the students.

A	B	C	D	E	F	G	H	I	J	K	L	M
3	1	1/4	2	1/2	3/4	1/2	1/4	1	2	1/2	3	3/4

In this activity, students are to find the point values for given words. Go over the examples below with the students. Then make copies of pages 47 and 48 and have the students practice. On page 48 is the activity "Super Frac-Code." This activity is more challenging than the one on page 47. Students can complete it individually or in pairs. You might want to go over the fraction and decimal equivalents before students begin the activity.

Examples:

L A D
3 3 2

Students are to add the numbers to get the point value for the word. In this example, the point value is 8.

H A M
1/4 3 3/4

Here, the point value is 4.

C A B
1/4 3 1

Here, the point value is 4 1/4.

M A K E
3/4 3 1/2 1/2

This word's point value is 4 3/4.

F I L L E D
3/4 1 3 3 1/2 2

Here, the point value is 10 1/4.

Point out to the students that the words LAD and HAM are words that are termed "wholesome." That is, they are words that do not end with a fraction.

Name ____________________ Date ____________

everyday problem solving

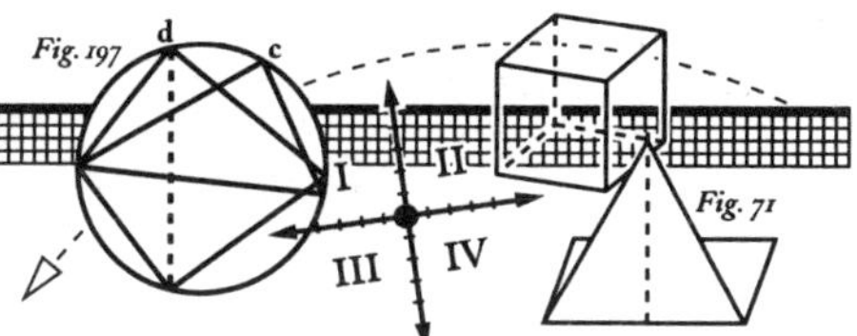

FRAC-CODE CONTINUED

Extensions:

- In a ten-minute time period, have students record all the words they can that have a point value greater than 6.125.
- In a ten-minute time period, have students record "wholesome" words.
- Have students find the Frac-Code point value for each of the 50 states. Ask them how many are "wholesome."
- Have students find the Frac-Code point value for the names of class members, as well as for famous people in history.
- Students can use the Frac-Code to organize this week's spelling list arranging the words from the highest point value to the lowest point value.

Name ______________________ Date ______________

everyday problem solving

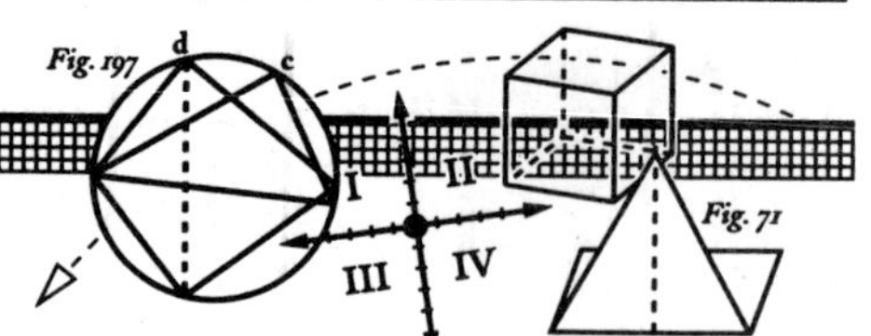

FRAC-CODE

Find the point value for the words below. Circle all of the "wholesome" words.

A	B	C	D	E	F	G	H	I	J	K	L	M
3	1	1/4	2	1/2	3/4	1/2	1/4	1	2	1/2	3	3/4

N	O	P	Q	R	S	T	U	V	W	X	Y	Z
1/2	1/4	1	1/4	3/4	1/2	3/4	2	1/4	3	1/4	1	2

1. M I L K ______
2. C A M E ______
3. H A L L ______
4. L A D L E ______
5. C H A L K ______
6. C A L M ______
7. F L A M E ______
8. B E A C H ______
9. L I F E ______
10. C A B L E ______
11. C A G E D ______
12. H A I L ______
13. F A B L E ______
14. F L A G ______
15. C H I D E ______
16. D E F A M E ______
17. R U S T Y ______
18. W O R R Y ______
19. W O R N ______
20. S T U N ______
21. T O W N ______
22. S N O W Y ______
23. S T O P ______
24. V O W S ______
25. S T O W ______
26. N O O N ______

Name ______________________________ Date ______________

everyday problem solving

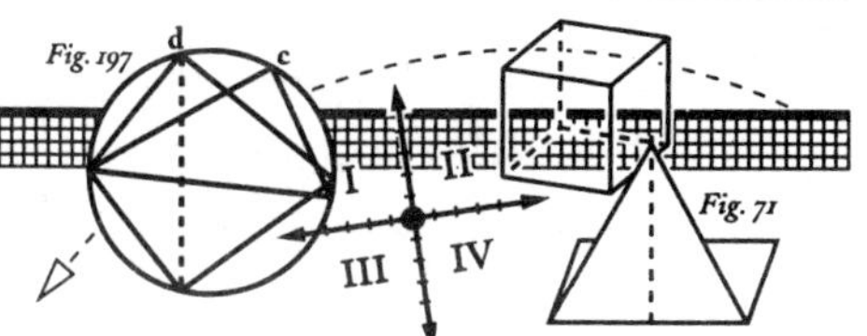

SUPER FRAC-CODE

All of the answers for the Super Frac-Code should be written in decimal notation. Find the decimal equivalent for each of the fractions to the right. This chart will help you when recording the answers.

Examples:

P	O	S	R	
1	1/4	1/2	3/4	Point value is 2.5.

S	T	O	R	Y	
1/2	3/4	1/4	3/4	1	Point value is 3.25.

7/8 = ______________

5/8 = ______________

3/8 = ______________

1/8 = ______________

3/4 = ______________

1/2 = ______________

1/4 = ______________

A	B	C	D	E	F	G	H	I	J	K	L	M
1/2	1/4	1	1/8	1/2	3/4	5/8	2	1/8	1/2	3	1/2	2

N	O	P	Q	R	S	T	U	V	W	X	Y	Z
1/8	3	1/8	1/2	2	1/8	2	1	3/4	1/8	1/8	3	3/4

1. A U T O ______
2. A N G L E ______
3. M A R C H ______
4. P I C T U R E ______
5. O C E A N ______
6. B U I L D I N G ______
7. T H R O W ______
8. T O S S E D ______
9. W O R K S ______
10. T O M A T O ______
11. P A C K A G E ______
12. D E C I S I O N ______
13. C R U N C H ______
14. I N V E S T I G A T E ______

Name____ Date ______________

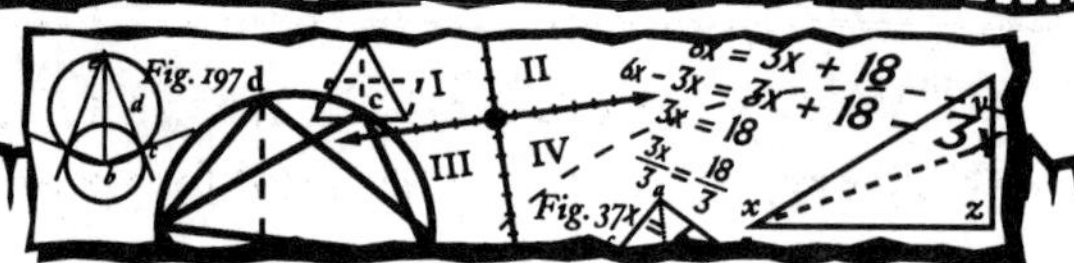

TIME IS EVERYTHING

Materials: copies of pages 50 and 51

Directions: Each location point on the perimeter of the square below has been given a specific number value. When attempting to decode a positional value, it will be helpful to think about the face of a clock.

Example:

11 12 1

10 2

9 3

8 4

7 6 5

= 6

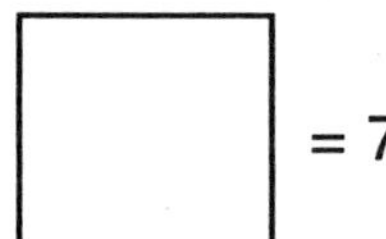

= 7

= 3

In this activity, the locations are used to solve math problems. Put the examples below on the board. Go over each one with the students. Then give students copies of pages 50 and 51 to complete.

Extensions: Have students make up their own problems and exchange them with classmates to solve.

12 x 2 = 24

9 x 3 = 27

11 x 7 = 77

(+) x = 44

(+) x (+) = 84

Name ______________________ Date ____________

everyday **problem solving**

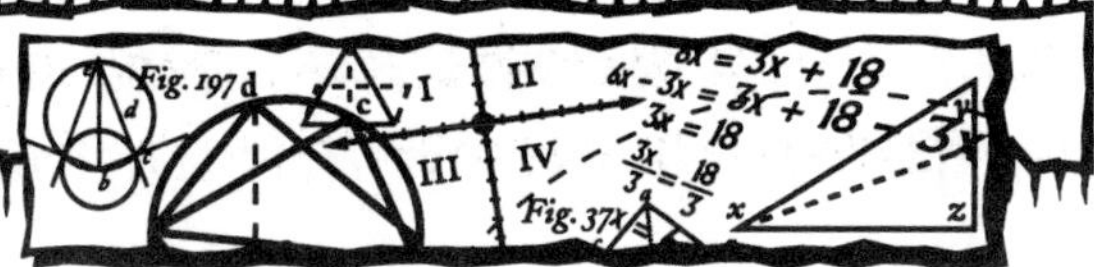

TIME IS EVERYTHING

Solve the problems below and on page 51.
Hint: Think about a clock face.

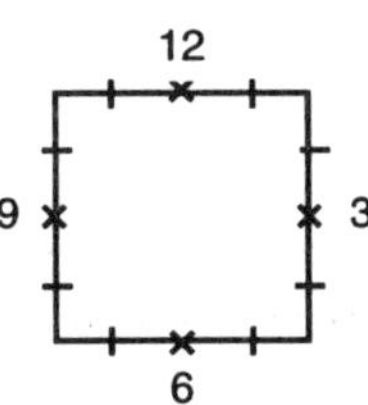

1. 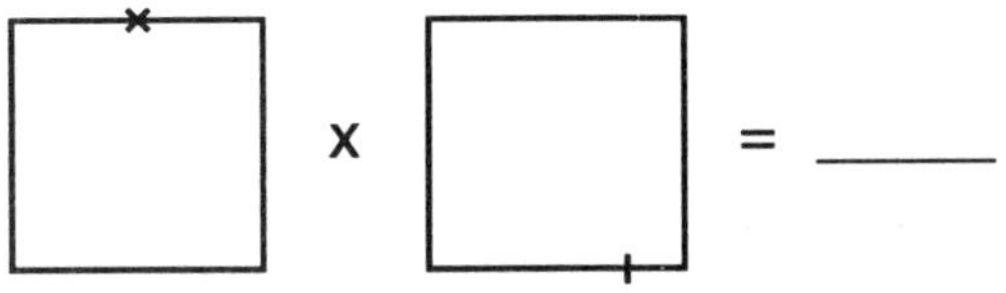 [ ] x [ ] = ______

2. 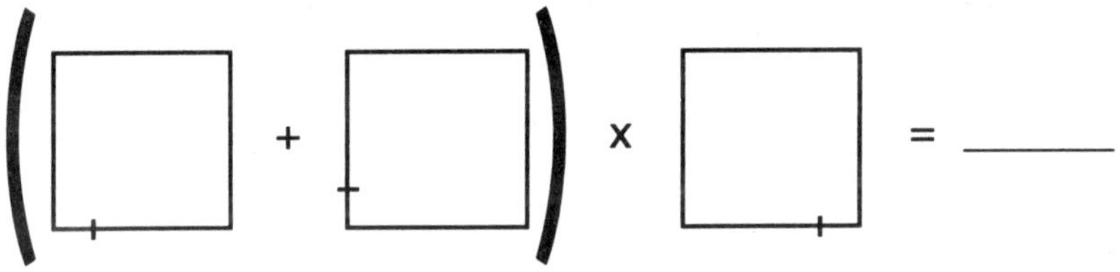 ([ ] + [ ]) x [ ] = ______

3. 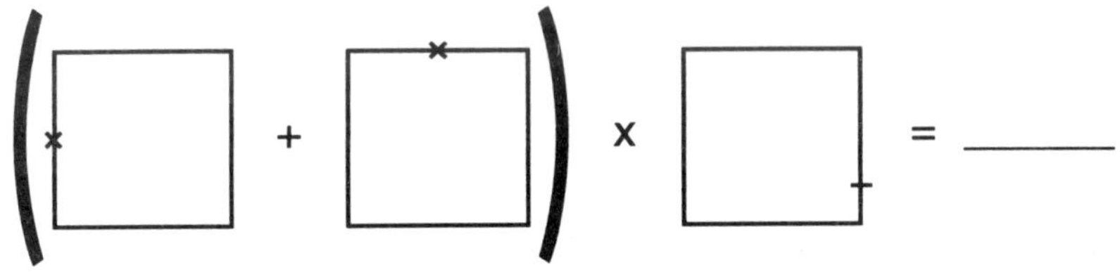 ([ ] + [ ]) x [ ] = ______

4. 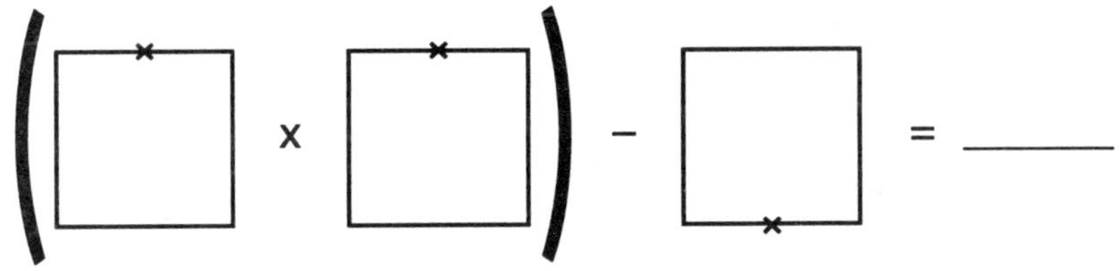 ([ ] x [ ]) – [ ] = ______

5. 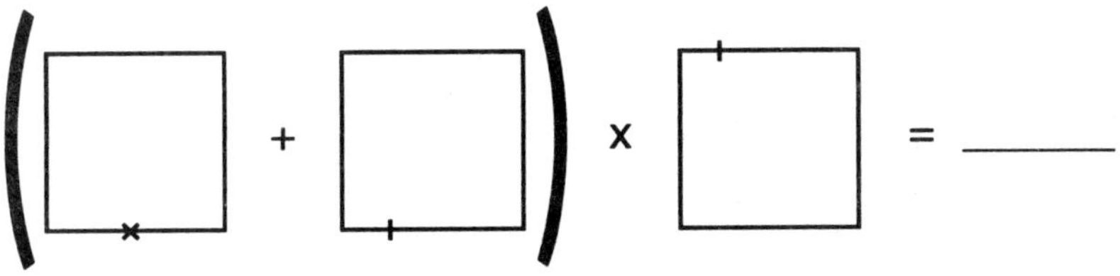 ([ ] + [ ]) x [ ] = ______

Name ____________________ Date ____________

everyday **problem solving**

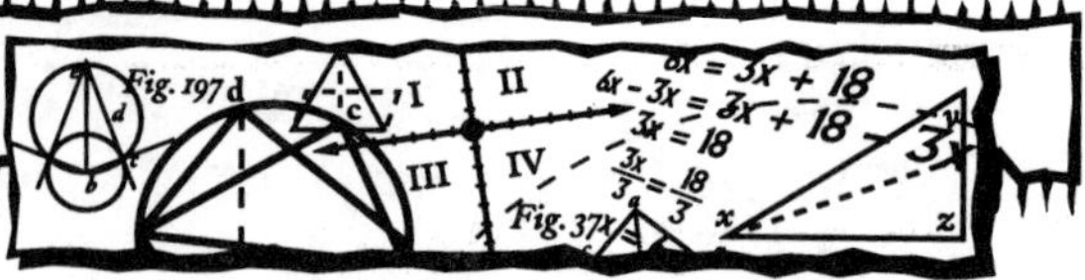

TIME IS EVERYTHING CONTINUED

6.

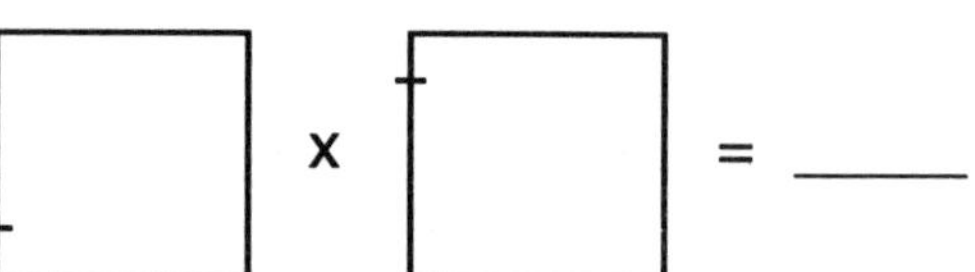

7. (☐ x ☐) x ☐ = ______

8. ☐ x (☐ + ☐) = ______

9. (☐ ÷ ☐) x ☐ = ______

10. ☐ + ☐ + ☐ = ______

11. (☐ x ☐) ÷ (☐ + ☐) = ______

Name______________________________ Date ______________

CODE ONE

Materials: copies of pages 53 and 54

Directions: Code One is a voice-activated number system. This system is an example of the binary system. Development of the system begins with the word PAT and continues with the words below. Each word is then assigned its own binary value.

PAT	**= 1**	**POT**	**= 8**
PET	**= 2**	**PUT**	**= 16**
PIT	**= 4**	**CAT**	**= 32**

Combining the six known words and their particular values with the addition operation, additional numbers can be created. Below are the first 10 numbers in the Code One number system and the corresponding values in the base ten system.

PAT	**= 1**	**PIT, PET**	**= 6**
PET	**= 2**	**PIT, PET, PAT**	**= 7**
PET, PAT	**= 3**	**POT**	**= 8**
PIT	**= 4**	**POT, PAT**	**= 9**
PIT, PAT	**= 5**	**POT, PET**	**= 10**

Go over the Code One system with the students. Then put the examples below on the board. Have students decode these numbers into the base ten system. Once students are comfortable with the system, give them copies of pages 53 and 54 to complete.

1. POT, PIT, PET = 8 + 4 + 2 = 14
2. PUT, PAT = 16 + 1 = 17
3. PUT, POT, PAT = 16 + 8 + 1 = 25
4. PUT, PET, PAT = 16 + 2 + 1 = 19
5. POT, PIT, PET, PAT = 8 + 4 + 2 + 1 = 15

Name ______________________ Date ______________

EVERYDAY PROBLEM SOLVING

CODE ONE

Decode the Code One numbers below and on page 54 to the base ten system.

PAT	= 1	POT	= 8
PET	= 2	PUT	= 16
PIT	= 4	CAT	= 32

1. POT, PIT ________
2. PIT, PET ________
3. POT, PAT ________
4. PUT, PET, PAT ________
5. CAT, PET ________
6. CAT, PIT, PET ________
7. CAT, POT, PAT ________
8. PUT, PAT ________
9. PUT, POT, PAT ________
10. PUT, POT, PET, PAT ________
11. CAT, PUT, PET ________
12. PUT, POT, PIT ________
13. CAT, POT ________
14. CAT, POT, PIT ________
15. PIT, PET, PAT ________

Name ____________________ Date ____________

CODE ONE CONTINUED

16. CAT, PUT, POT ________

17. CAT, PAT ________

18. CAT, POT, PIT, PET ________

19. PUT, POT, PIT, PET, PAT ________

20. CAT, PUT, POT, PIT ________

Write the numbers below using the Code One system.

21. 20 ____________________

22. 13 ____________________

23. 34 ____________________

24. 37 ____________________

25. 41 ____________________

26. 38 ____________________

27. 49 ____________________

28. 52 ____________________

29. 27 ____________________

30. 55 ____________________

31. With the information you now have about Code One, what is the largest number that you can create? ________

Name ______________________________ Date ______________

everyday problem solving 

NEXT TO LAST PLACE WINS

Materials: four number cubes (numbered 1–6)

Directions: This activity is for two players or for two teams with up to five players on each team. The winning player will be the one in sole possession of Next to Last Place at the conclusion of the game.

Player One begins the game by tossing the four cubes. The top face on each of the four cubes will be the four numbers that Player One must use to create an equation. All four numbers must be used in the equation, and each number is to be used only once.

Example: Player One tosses 3, 6, 2, and 6. To create an equation with these four numbers, Player One may use any operation or combine operations to arrive at an answer. (6 x 2) + (6 x 3) = 30. Player One's number is 30. Player One turns over the four cubes to Player Two.

Player Two's task is to toss the four cubes and create his or her own equation. Player Two's four cubes are 6, 1, 4, and 5, and the equation is (4 + 1) x (6 + 5) = 55.

After the first round of play, last place, 30, belongs to Player One, and first place, 55, belongs to Player Two.

Round Two

Player One:	1, 4, 1, 2	4 x (2 + 1 + 1) = 16
Player Two:	5, 5, 2, 4	(5 x 5) + (4 – 2) = 27

The standings after the first two rounds of play are as follows:

55	Player Two
30	Player One
27	Player Two
16	Player One

At this point, Next to Last Place belongs to Player Two with a score of 27. Play continues in this manner with each player taking alternate turns and striving for that coveted Next to Last Place. The activity will last for 10 rounds of play.

Name________________________________ Date____________

SQUARES AND TRIANGLES

Materials: two number cubes (numbered 1–6), copies of page 58

Directions: This activity is for two players or two teams. (The class can be divided into two large teams.) While playing this game, the students will become aware of and recognize numbers that create squares and triangles.

Square Numbers

```
x x     x x x     x x x x
x x     x x x     x x x x
        x x x     x x x x
                  x x x x
 4        9         16
```

Triangular Numbers

```
x x     x x x     x x x x
x       x x       x x x
        x         x x
                  x
 3        6         10
```

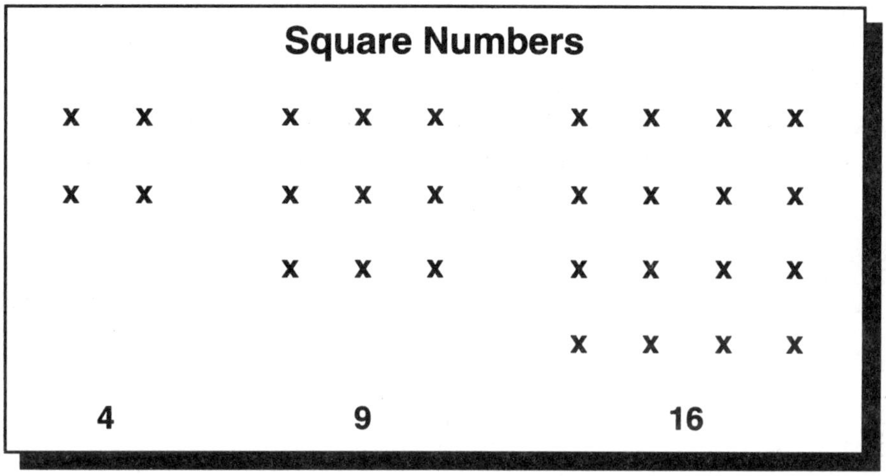

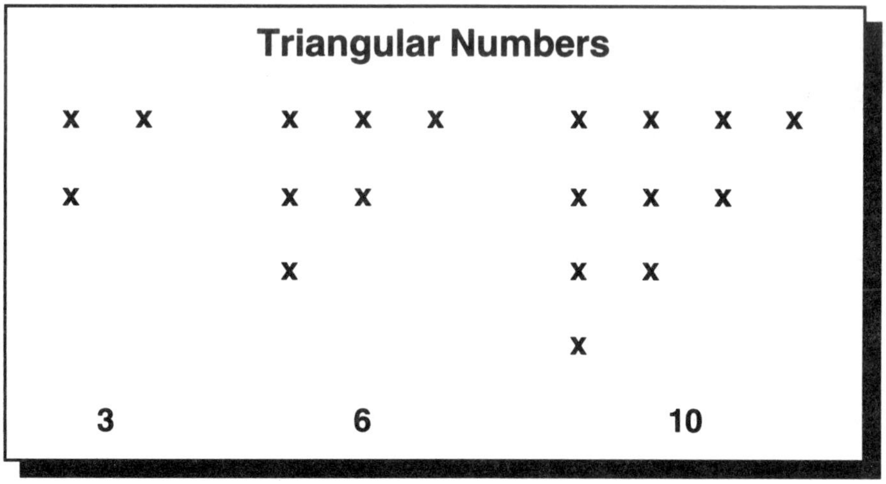

The objective of the game is to come as close to 100 as possible without exceeding 100 points. If no one is able to reach 100, then the player with a sum closest to 100 will be the winner.

everyday problem solving

Name____ Date______________

SQUARES AND TRIANGLES CONTINUED

Each player is allowed up to seven tosses with the cubes. The seven tosses constitute a turn. The players add the numbers on the faces of the cubes after each toss and move that many spaces on the gameboard (page 58). It is the player's choice as to when he or she chooses to stop the turn. A player does not have to use all seven tosses. The player might reach a point where he or she is just about as close to 100 as he or she can possibly be. No one is allowed to exceed the score of 100. To do so will automatically cause the player to lose the game.

Give each player a copy of page 58 to use to keep track of his or her score. All players begin at 0. If a player tosses the cubes and the sum allows the player to stop on a square number, then that player must go back two steps. If the sum of the cubes causes a player to stop on a triangular number, then the player is to advance two steps.

There is one number that all players will want to avoid if at all possible and that is the number 36. If tossing the cubes gives a player a sum that takes him or her to this point, the player is to return to 0 and begin play all over again. As you can see, the number 36 is both a square and a triangular number.

everyday problem solving

Name________________________________ Date______________

Squares and Triangles Gameboard

0	1	2	3	4	5	6	7	8	9	10
	11	12	13	14	15	16	17	18	19	20
	21	22	23	24	25	26	27	28	29	30
	31	32	33	34	35	36	37	38	39	40
	41	42	43	44	45	46	47	48	49	50
	51	52	53	54	55	56	57	58	59	60
	61	62	63	64	65	66	67	68	69	70
	71	72	73	74	75	76	77	78	79	80
	81	82	83	84	85	86	87	88	89	90
	91	92	93	94	95	96	97	98	99	100

Name ______________________ Date ______________

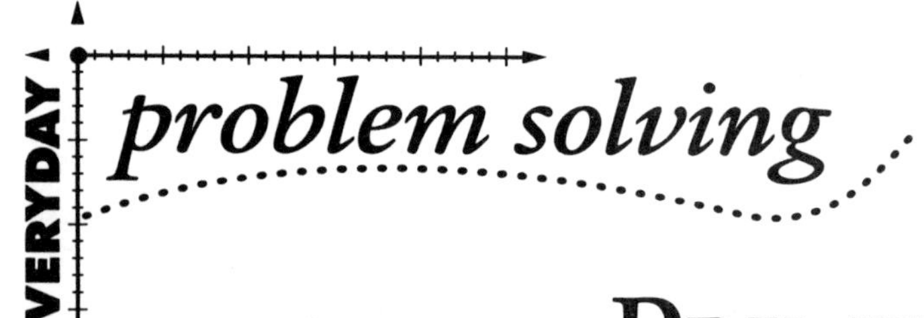

Prime Cubes

Materials: three prime number cubes (each cube should contain the numbers 2, 3, 5, 7, 11, and 13), 20 small markers, copies of pages 61–63

Directions: There are three different prime cube games in this activity. Each game is for two players or two teams with up to four players each. The objective of all three games is to have the lowest sum of all the remaining uncovered numbers on the gameboard (pages 61–63).

Prime Cube 2 Game

Give each player a copy of the Prime Cube 2 Gameboard (page 61) and 10 small markers. Players take turns tossing two of the prime number cubes. The sum of the cubes is found by adding the top face of each cube. Once this sum is determined, the player puts a marker on that number space on his or her gameboard. Each player will toss the two prime cubes as many times as needed to cover 10 of the gameboard number spaces.

When 10 number spaces are covered, the remaining uncovered spaces are to be totaled. This sum is the player's score. The lowest sum recorded by all players is the winner of the game.

Prime Cube 3 Game

The Prime Cube 3 Game is open to many possible answers because students can use all three prime number cubes, and they are allowed to use and/or combine the four basic operations.

Two players or three teams with up to five players on each team can play this game. Players take turns tossing the three cubes. The top number on each cube may be used in any basic operation or in a combination of operations to attain any of the answers on the Prime Cube 3 Gameboard (page 62). The numbers displayed on the face of each cube may only be used one time each. Each player keeps a running total of his or her points. (The number of points is the answer that the player got on the gameboard.) Each player will be allowed to toss the cubes seven times. The objective of the game is to accumulate the most points after the seventh toss of the cubes.

Name ______________________ Date ____________

problem solving

PRIME CUBES CONTINUED

Prime Cube 3390 Game

This version of Prime Cubes is somewhat different from the two previous games. This game is for two to four players or three teams with up to four players on each team.

Each player begins with a score of 3390. Players take turns tossing three prime cubes. Players may perform any operation or combine operations on all three cubes to create any of the numbers seen on the Prime Cube 3390 Gameboard (page 63). Each player subtracts the answer to his or her equation (the number on the gameboard) from 3390. The objective of the game is to have the lowest score.

Example:

Player A tosses 13, 3, and 5. After some careful deliberation, the player may record the equation (13 x 5) x 3 = 195. Since this answer is on the gameboard, the player owns this number and subtracts it from 3390. (3390 - 195 = 3195) At this point, the number 195 is no longer available to the player.

Player A's second toss of the cubes results in 13, 5, and 11. The equation is (5 + 11) x 13 = 208. This answer is available on the gameboard. This number should then be subtracted from 3195 (3195 - 208 = 2987). The lower the final score, the better the position is for the player.

If for some reason an equation cannot be created that will match an answer on the gameboard, the player loses one of his or her 10 cube tosses.

Name________________________ Date ____________

EVERYDAY *problem solving*

PRIME CUBE 2 GAMEBOARD

4	**13**	**20**	**14**	**5**
8	**13**	**32**	**16**	**10**
35	**41**		**60**	**37**
9	**14**	**33**	**18**	**12**
6	**15**	**22**	**16**	**7**

Name ______________________ Date ______________

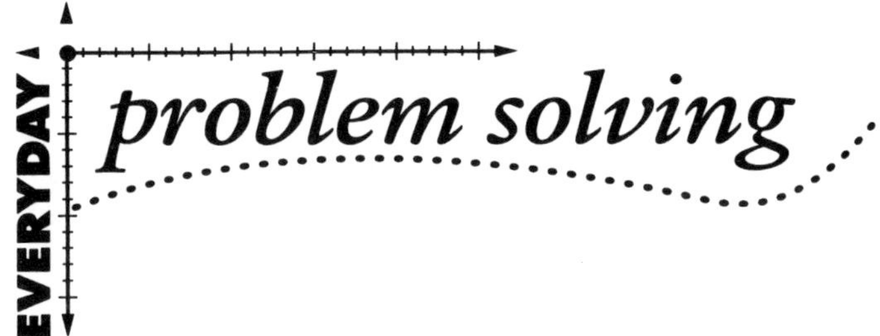

Prime Cube 3 Gameboard

1	6	7	15	20	21
22	25	27	28	30	33
36	39	40	48	52	66
70	78	88	90	100	105
120	125	130	175	220	275

Name ______________________ Date ______________

EVERYDAY *problem solving*

PRIME CUBE 3390 GAMEBOARD

385	**273**	**220**	**208**	**195**
182	**168**	**165**	**156**	**154**
150	**130**	**112**	**110**	**105**
100	**90**	**76**	**72**	**70**
68	**66**	**52**	**48**	**35**

Name________________________________ Date ______________

everyday problem solving

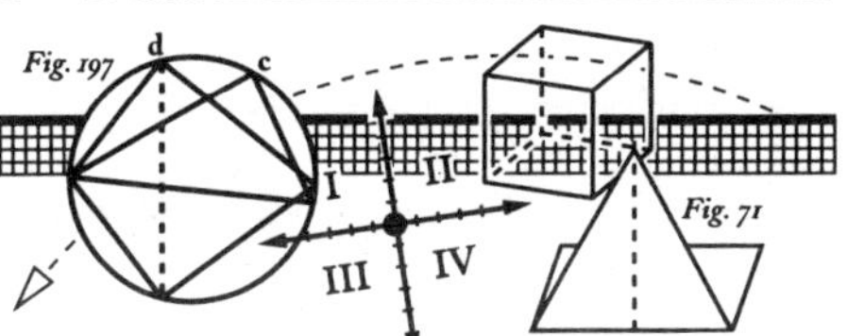

BUY A LETTER

Materials: two number cubes (six faces on one cube should be labeled 1–6, six faces on the other cube should be labeled 4–9), copies of pages 65–67

Directions: This activity is for two players or three teams with up to three players on each team. Players are to toss the two number cubes and multiply the numbers that appear on the top face of each cube. Whatever the product is, the players should be able to buy a letter from the given words (on top of pages 65–67). The cost of the letter is written beneath each one. The players may only use the letters in the words that are offered for sale on each page. After buying a letter, the players record their purchase on the activity page.

The goal is to be able to have the largest possible score through the purchase of letters and the construction of the words. This total will determine a player's score. Each round will allow each player to toss the number cubes three times. Players will take turns tossing the cubes. The largest total score at the end of the time limit (15 minutes) will determine the game winner.

Example: The activity on page 65 offers for sale the letters in the two words *Problem Solving.*

P	R	O	B	L	E	M		S	O	L	V	I	N	G
58	48	45	42	40	32	28		27	45	40	20	18	10	4

After the first round, Player A was able to buy the letters I, M, and L. At the start of the next round, Player A was able to buy the letter E. Player A may now record the word LIME, and its point value is 118. (L = 40, I = 18, M = 28, E = 32, total = 118) Remember, the winning player will have the largest point value total and not necessarily the most words. Once a letter is bought, it can be used as many times as necessary when making words. However, each letter can only be used once in each word, but a player may buy two of the same letters. If a player does buy two of the same letters, this player may use both of the letters in the word (i.e., LOOP).

Name ______________________ Date ______________

everyday problem solving

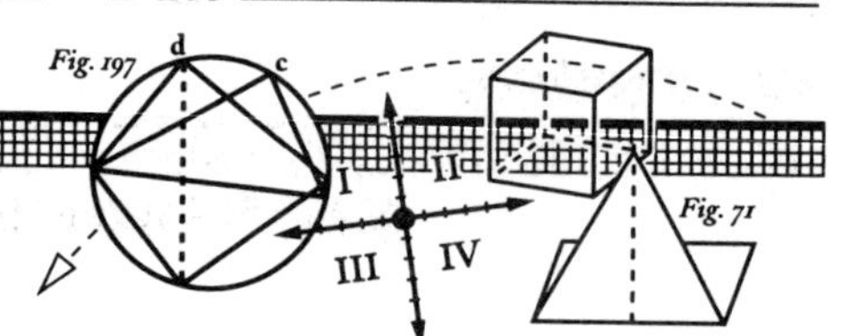

BUY A LETTER—
PROBLEM SOLVING

Circle all of the letters that you bought. After buying letters, try to create at least 25 words within the time limit. Remember to find the point value of the words.

P	R	O	B	L	E	M		S	O	L	V	I	N	G
58	48	45	42	40	32	28		27	45	40	20	18	10	4

1. ______________________
2. ______________________
3. ______________________
4. ______________________
5. ______________________
6. ______________________
7. ______________________
8. ______________________
9. ______________________
10. ______________________
11. ______________________
12. ______________________
13. ______________________
14. ______________________
15. ______________________
16. ______________________
17. ______________________
18. ______________________
19. ______________________
20. ______________________
21. ______________________
22. ______________________
23. ______________________
24. ______________________
25. ______________________

VANNA!
VANNA!
VANNA!

Name ______________________________ Date ______________

everyday

problem solving

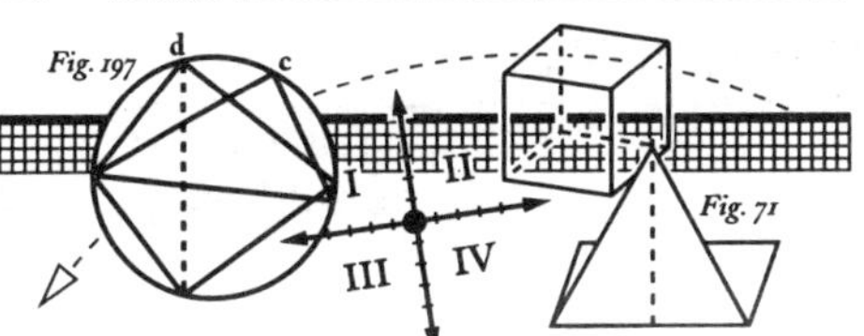

Buy a Letter—
Mathemactivities

Circle all of the letters that you bought. After buying letters, try to create at least 25 words within the time limit. Remember to find the point value of the words.

M	A	T	H	E	M	A	C	T	I	V	I	T	I	E	S
54	48	45	42	40	54	48	32	45	28	10	28	45	28	40	4

1. ______________________
2. ______________________
3. ______________________
4. ______________________
5. ______________________
6. ______________________
7. ______________________
8. ______________________
9. ______________________
10. ______________________
11. ______________________
12. ______________________
13. ______________________
14. ______________________
15. ______________________
16. ______________________
17. ______________________
18. ______________________
19. ______________________
20. ______________________
21. ______________________
22. ______________________
23. ______________________
24. ______________________
25. ______________________

I'D LIKE TO BUY A VOWEL!

Name ______________________ Date ______________

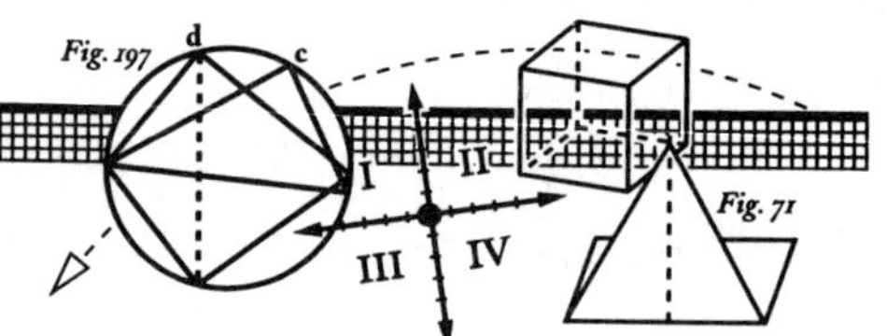

BUY A LETTER—
INVESTIGATIONS

Circle all of the letters that you bought. After buying letters, try to create at least 25 words within the time limit. Remember to find the point value of the words.

I	N	V	E	S	T	I	G	A	T	I	O	N	S
4	40	28	45	32	42	4	54	48	42	4	36	40	32

1. ______________________
2. ______________________
3. ______________________
4. ______________________
5. ______________________
6. ______________________
7. ______________________
8. ______________________
9. ______________________
10. ______________________
11. ______________________
12. ______________________
13. ______________________
14. ______________________
15. ______________________
16. ______________________
17. ______________________
18. ______________________
19. ______________________
20. ______________________
21. ______________________
22. ______________________
23. ______________________
24. ______________________
25. ______________________

MA'AM, CAN YOU DESCRIBE THE LETTER AGAIN?

Name ______________________ Date ______________

everyday **problem solving**

BLACK AND RED

Materials: two cubes (numbered 1–6), 42 playing cards (all aces through 10s and two jokers)

Directions: Black and Red can be played by two players or two teams with two players on each team. The playing cards are to be accepted at face value. (The ace is equivalent to 1.) The four suits determine the operation. (See box below.) The objective of the game is to have the largest even number at the end of the game. The activity allows seven rounds of play.

♠ **spades = addition**

♣ **clubs = multiplication**

♦ **diamonds = subtraction**

♥ **hearts = division**

Example: Player A tosses the number cubes. On the top face of one cube is a 3. The other cube shows a 6. The number is recorded as 36 or 63. The choice should be 63.

At this juncture, the 42 cards are shuffled and placed facedown. The top card, a nine of clubs, is drawn. Player A would record 63 x 9 = 567. For round two, Player A draws a three of hearts. This creates the equation 567 ÷ 3 = 189. For round three, Player A draws a five of hearts which creates the equation 189 ÷ 5 = 37.8. (When a decimal is formed, it should be rounded to a whole number. In this case, 38.) For round four, Player A draws a seven of clubs and the equation 38 x 7 = 266 is created.

Play continues in this manner until Player A has completed seven rounds. Hopefully, at the end of the seventh round, the final answer will be larger than that of Player B, and it will also be an even number.

There is one final and most important rule. If during the course of play, a player is unfortunate enough to expose one of the two jokers, the score for him or her will revert back to zero, and the player must begin to rebuild his or her score by starting with the unused rounds.

Name ______________________ Date ______________

PRIMES OUT

Materials: two number cubes (numbered 1–6)

Directions: This activity is for two players, or the class can be divided into two teams. The objective is for a player to eliminate as many of the first 11 prime numbers as possible. (The first 11 prime numbers are 2, 3, 5, 7, 11, 13, 17, 19, 23, 29, and 31.) In order to accomplish this feat, each team will be given 25 chances at tossing the cubes. After tossing the cubes, the numbers on the top face of each cube will be added. This sum is then used to match and then erase a prime number.

At the end of the game, the player with the lowest score is the winner. All of the prime numbers that the player could not erase will be added, and this sum will be compared to the total number of prime numbers the other player could not erase.

Example: Follow this play through the first seven tosses by Player A.

a. **6 + 4 = 10 should be credited to prime 31.**

b. **6 + 5 = 11 allows for the elimination of the prime 11.**

c. **2 + 3 = 5 allows for the elimination of the prime 5.**

d. **5 + 4 = 9 should be credited to prime 19.**

e. **4 + 6 = 10 should be credited to prime 19, and this will eliminate the prime 19.**

f. **5 + 5 = 10 should be credited to prime 31 (for a total of 20).**

g. **3 + 1 = 4 should be credited to prime 31 (for a total of 24).**

After seven rounds, this team has eliminated the prime numbers 5, 11, and 19.

With 18 tosses still available to Player A, the first time that he or she tosses a 7, this number should be credited to the prime 31 and will allow for its elimination from the game.

After 25 tosses of the cubes, prime numbers not eliminated are to be added. This sum is to be compared with the sum compiled by the other players. The player with the lowest sum is the winner.

Name ______________________ Date ______________

everyday **problem solving** 

YOUR GUESS!

Materials: five number cubes (numbered 1–6), copies of the score sheet (below)

Directions: This game is for two players. The game will consist of five rounds of play. Each player will complete his or her round of play before the numbered cubes are passed to the next player.

Initial play begins with Player One getting ready to toss the five cubes. Before tossing the cubes, Player One is to estimate the sum he or she thinks will appear on the cubes.

Before the cubes are thrown, Player One will record this estimate on the score sheet. After the cubes are thrown, the player will record the actual sum of the five cubes.

During the second round, Player One does the same procedure except he or she will only be using four cubes. This format continues until Player One is finally tossing only one cube. Player One then passes the cubes to Player Two.

The objective of Your Guess! is to have the lowest score after both players have completed the five rounds. This score is to be tallied by adding the five recorded estimates and then finding the sum of the five actual tosses. Subtract the smaller number from the larger number, and the players will arrive at a difference. The winner is the player with the smallest difference after all five rounds.

Round	Estimate	Actual Sum	Difference
five cubes			
four cubes			
three cubes			
two cubes			
one cube			
totals			
		Final Difference	

Name________________________ Date__________

999

Materials: six number cubes (numbered 1–6)

Directions: This game is for two players or two teams. The objective is to have the largest score closest to 999 without exceeding this number. The game continues for five rounds for each player. In turn, each player will toss the six number cubes and place the cubes in numerical order. The next step is to group the three largest number cubes and the three smallest number cubes. The cubes should then be arranged in each set so that they form the largest three-digit number possible for that particular set. Once both sets of three-digit numbers have been established, the idea is then to subtract the smaller number from the larger one.

Example: Player A tossed the six number cubes. They landed on the numbers shown below.

5, 2, 4, 4, 5, 6

The two sets of numbers are 6, 5, 5 and 4, 4, 2. Therefore, the two numbers are 655 and 442. After subtracting the numbers, the difference is 213. This is the initial score for Player A. Player A is now on the way to reaching the 999 total.

Player A now has up to four more opportunities to try to reach a score as close to 999 without exceeding it. At any time, a player may decide that his or her score is as close to 999 as it will ever be and choose to stop. At this point, the player accepts his or her score and gives the six cubes to the next player. Player B now tries to surpass Player A's score, but he or she must remember not to go over 999.

Name________________________________ Date____________

999 CONTINUED

Example of six cube tosses for Player A:

six cube tosses		arrangement of numbers	score
first round	5, 2, 4, 4, 5, 6	655 – 442 = 213	213
second round	3, 1, 5, 1, 3, 6	653 – 311 = 342 + 213 = 555	555
third round	4, 1, 2, 1, 3, 2	432 – 211 = 221 + 555 = 776	776
fourth round	5, 6, 1, 1, 6, 5	665 – 511 = 154 + 776 = 930	930

At this juncture, Player A accepts his or her score of 930 and decides not to exercise the fifth round.

The six cubes are given to Player B. It is now the goal of Player B to toss the cubes for up to five rounds with the hope of surpassing the 930 score by Player A but also keeping in mind that he or she can at no time exceed the score of 999.

Name ______________________ Date ____________

NYSE (New York Stock Exchange)

Materials: one hundred 2 ½" x 3 ½" cards

An efficient way of constructing the cards would be to divide the 100 blank cards among groups of students and then assign one or two companies to each group. (Companies are listed on page 74.) It would be up to students to list the company information on each card. There are 20 companies, and each company will have five identical cards.

Each card should contain the following information:

- Company Name
- NYSE Ticker Symbol
- Share Price
- Brief Business Description

BUY LOW! SELL HIGH!

TICK! TICK! TICK!

Sample Cards:

Disney	Coca-Cola	Mobil	Time Warner
DIS	KO	MOB	TWX
82	63	131	45
Amusement Parks, Films	Soft Drinks	Oil	Publishing, Entertainment

Name ______________________ Date ______________

problem solving

NYSE (New York Stock Exchange) continued

Company Information Sheet

Company Name	Ticker Symbol	NYSE Share Price	Business Description
American Telegraph and Telephone	T	33	Telecommunications
Chrysler	C	30	Auto Maker
Coca-Cola	KO	63	Soft Drinks
Disney	DIS	82	Amusement Parks, Films
Ford	F	35	Auto Maker
General Dynamics	GD	71	Defense Products
General Electric	GE	111	Consumer Products
General Motors	GM	58	Auto Maker
Hershey Foods	HSY	54	Candy
Home Depot	HD	58	Home Building Products
International Business Machines	IBM	161	Computers
Johnson & Johnson	JNJ	61	Health Care Products
McDonalds	MCD	54	Fast-Food Restaurant
Mobil	MOB	131	Oil
Pepsi Cola	PEP	34	Soft Drinks
Reebok	RBK	38	Shoes, Apparel
Sears	S	49	Large Retailer
Sony	SNE	73	Electronics
Time Warner	TWX	45	Publishing, Entertainment
WalMart	WMT	28	Discount Stores

Name ______________________ Date ______________

EVERYDAY *problem solving*

NYSE (New York Stock Exchange) continued

Directions: This game is for two teams with up to five players on each team. All 100 company cards are to be shuffled. Each team is to receive 10 cards. The players will hold the cards in their hands so as not to allow the opposing team the opportunity of knowing what they are. The remaining 80 cards will be stacked in an unexposed pile.

The players are to regard each card as a share in that company's stock. When a team is able to collect three or more shares of a particular stock, they are then entitled to own them outright. The players are to remove these share cards from their hands and display them faceup on the playing surface for all to see.

From the unexposed deck, the team is to draw cards so that it always has 10 cards. Ownership in a company is exposing three, four, or five cards in that company. Teams will maintain 10 cards in their hands as long as possible.

There may be a time in the game when a team holding 10 share cards will not have three cards of the same company. At this point, it will draw the top card from the unexposed deck and if it decides to keep the card, the team must discard one of its previously held share cards in another pile facedown.

Now it is the opposing team's turn to draw a card, and it has the option to select the share card that was just discarded, or it may elect to take the next share card from the unexposed deck.

If a team has shown three share cards from the same company, and if the team is fortunate enough to pull another share card from the same company, it may immediately add that share card to its exposed collection.

Name ____________________ Date ____________

NYSE (New York Stock Exchange) continued

Example: Team A is showing three shares of Ford. It is able to draw a fourth share of Ford, so it now shows four exposed shares of Ford. A team member may then draw an additional share card from the deck to maintain its total of 10 share cards.

At the end of the game, the team with the highest combined company share card value will be declared the game winner.

For example, perhaps at the end of the game, Team A's holdings were as follows:

4 shares of Disney

5 shares of McDonalds

3 shares of General Motors

5 shares of Ford

The total would be this:

DIS	4 x 82 =	328
MCD	5 x 54 =	270
GM	3 x 58 =	174
F	5 x 35 =	175
Total		**$947.00**

Note: As the share cards are being constructed, a good idea might be to cover the part of the share card that will indicate the price of the card with clear transparent paper so that a wax pencil can be used to record the current share price.

Share card prices are rounded off to the nearest dollar.

ANSWERS

Page 9

a. 121 = May 1
b. 59 = February 28
c. 200 = July 19
d. 156 = June 5
e. 135 = May 15
f. 222 = August 10
g. 277 = October 4
h. 340 = December 6
i. 322 = November 18
j. 257 = September 14
k. 195 = July 14
l. 100 = April 10
m. 246 = September 3
n. 47 = February 16
o. 282 = October 9
p. 300 = October 27
q. 93 = April 3
r. 139 = May 19
s. 238 = August 26
t. 95 = April 5
u. 553 = July 7
v. 707 = December 8
w. 392 = January 27
x. 648 = October 10
y. 421 = February 25
z. 1005 = October 2

Page 12

1. 8/50 = 0.16
2. 8/50 = 0.16
3. 4/50 = 0.08
4. 3/50 = 0.06
5. 2/50 = 0.04
6. 2/50 = 0.04
7. 11/50 = 0.22
8. 3/50 = 0.06
9. 32/50 = 0.64
10. 10/50 = 0.2
11. 48/50 = 0.96
12. 9/50 = 0.18
13. 20/50 = 0.4
14. 24/50 = 0.48
15. 1/50 = 0.02

Page 21

a. Four-digit numbers that are less than 5000
b. Four-digit numbers that use the same digits
c. Four-digit numbers that are even
d. Three historical years
e. In each four-digit number, the tens and ones are one half of the thousands and hundreds.
f. Each four-digit number is comprised of consecutive digits.
g. Each four-digit number has a zero in the hundreds place.
h. The four-digit numbers are comprised of the same digits.
i. Double digits are adjacent in each of the four-digit numbers.
j. Each four-digit number is a palindrome.
k. There are six perfect square numbers (25, 64, 36, 49, 81, 16).
l. Each four-digit number begins and ends with the same number.
m. Each four-digit number is divisible by 3.
n. The sum of the digits in the hundreds, tens, and ones place will add up to the number in the thousands place.
o. The 12 digits that make up this problem are in even/odd order.
p. Each four-digit number has a digital root of 9.
q. Each four-digit number is divisible by 3.
r. Nine numbers which make up the beginning of the Fibonacci Sequence (1, 1, 2, 3, 5, 8, 13, 21, 34)
s. The sum of the four-digit numbers is 5000.

Page 31

a. 19 = bananas, hot fudge, ice cream
b. 35 = marshmallows, hot fudge, ice cream
c. 21 = bananas, caramel, ice cream
d. 37 = marshmallows, caramel, ice cream
e. 25 = bananas, nuts, ice cream
f. 13 = nuts, caramel, ice cream
g. 27 = bananas, nuts, hot fudge, ice cream
h. 41 = marshmallows, nuts, ice cream
i. 49 = marshmallows, bananas, ice cream
j. 23 = bananas, caramel, hot fudge, ice cream
k. 65 = sprinkles, ice cream

l. 51 = marshmallows, bananas, hot fudge, ice cream

m. 15 = nuts, caramel, hot fudge, ice cream

n. 31 = bananas, nuts, caramel, hot fudge, ice cream

o. 97 = sprinkles, marshmallows, ice cream

p. 33 = marshmallows, ice cream

q. 29 = bananas, nuts, caramel, ice cream

r. 101 = sprinkles, marshmallows, caramel, ice cream

The owner became upset with the five orders because the customers didn't order any ice cream.

Page 33

A. 17 days + 4 days = 3 weeks

B. 27 inches + 9 inches = 1 yard

C. 40 ounces + 8 ounces = 3 pounds

D. 725 days + 5 days = 2 years

E. 13 + 11 = 2 dozen

F. 119 + 25 = 1 gross

G. 1100 pounds + 900 pounds = 1 ton

H. 86 years + 14 years = 1 century

I. 15 years + 5 years = 2 decades

J. 33 inches + 15 inches = 4 feet

Pages 35–36

1. seven = hou**se ven**ture
2. two = ca**t wo**uld
3. six = her**s I x**-rated
4. one = wh**o ne**ed
5. ten = Wha**t en**tertainment
6. twelve = Ar**twel Ve**rnon
7. five = I**f I've**
8. ninety = Ca**n I net y**our
9. three = Bo**th ree**ds
10. four = i**f our** team
11. hundred = s**hun dred**ging
12. eleven = hot**el Even**ings
13. twenty = Scot**t went y**early
14. eight = n**eigh t**o
15. seventy = wa**s, even Ty**rone
16. fifty = Jef**f, if Ty**rell
17. nine = Ca**n I ne**ver
18. eighty = fr**eight y**ou
19. forty = Puf**f or Ty**ke

Page 38

a. 18 b. 20
c. 28 d. 32
e. 36 f. 55
g. 14 h. 35
i. 59 j. 24
k. 17 l. 81
m. 83 n. 85
o. 62 p. 84
q. 53 r. 52
s. 71 t. 242

Page 40

1. 16 = Randall Cunningham
2. 41 = Joe Namath
3. 13 = Bobby Clarke
4. 17 = Joe DiMaggio
5. 7 = Wade Boggs
6. 37 = Willie Mays
7. 56 = Emmitt Smith
8. 3 = Ernie Banks
9. 28 = Sandy Koufax
10. 43 = Shaquille O'Neal
11. 26 = Magic Johnson
12. 35 = Dan Marino
13. 6 = Drew Bledsoe
14. 11 = Steve Carlton
15. 60 = Reggie White
16. 38 = Joe Montana
17. 29 = Mario Lemieux
18. 62 = Ted Williams
19. 9 = Jim Brown
20. 5 = Larry Bird
21. 22 = Jimmy Foxx
22. 14 = Roberto Clemente
23. 61 = Dominique Wilkins
24. 39 = Warren Moon
25. A, B
26. B, D, F
27. A, B, D, F
28. A, C, D

Page 47

Bold numbers are "wholesome" numbers.

1. 5 ¼ 2. 4 ½
3. 9 ¼ 4. 11 ½
5. **7** 6. **7**
7. **8** 8. **5**
9. 5 ¼ 10. 7 ¾
11. 6 ¼ 12. 7 ¼
13. 8 ¼ 14. 7 ¼
15. **4** 16. 7 ½
17. **5** 18. 5 ¾
19. 4 ½ 20. 3 ¾
21. 4 ½ 22. 4 ¼
23. 2 ½ 24. **4**
25. 4 ½ 26. 1 ½

Page 48

$\frac{7}{8}$ = 0.875	$\frac{5}{8}$ = 0.625
$\frac{3}{8}$ = 0.375	$\frac{1}{8}$ = 0.125
$\frac{3}{4}$ = 0.75	$\frac{1}{2}$ = 0.5
$\frac{1}{4}$ = 0.25	

1. 6.5 **2.** 2.25
3. 7.5 **4.** 6.75
5. 5.125 **6.** 2.875
7. 9.125 **8.** 5.875
9. 8.25 **10.** 12.5
11. 6.25 **12.** 5.125
13. 7.125 **14.** 7.375

Pages 50–51

1. 12 x 5 = 60
2. (7 + 8) x 5 = 75
3. (9 + 12) x 4 = 84
4. (12 x 12) – 6 = 138
5. (6 + 7) x 11 = 143
6. 8 x 10 = 80
7. (9 x 11) x 4 = 396
8. 6 x (7 + 8) = 90
9. (12 ÷ 3) x 10 = 40
10. 10 + 11 + 12 = 33
11. (12 x 2) ÷ (5 + 1) = 4

Pages 53–54

1. 12 **2.** 6
3. 9 **4.** 19
5. 34 **6.** 38
7. 41 **8.** 17
9. 25 **10.** 27
11. 50 **12.** 28
13. 40 **14.** 44
15. 7 **16.** 56
17. 33 **18.** 46
19. 31 **20.** 60
21. PUT, PIT
22. POT, PIT, PAT
23. CAT, PET
24. CAT, PIT, PAT
25. CAT, POT, PAT
26. CAT, PIT, PET
27. CAT, PUT, PAT
28. CAT, PUT, PIT
29. PUT, POT, PET, PAT
30. CAT, PUT, PIT, PET, PAT
31. 63

Pages 65–67

Possible words:

Problem Solving
bill, boil, bring, broom, globe, glove, gone, goose, grim, groom, grove, live, loop, love, loving, mile, mobile, mole, moon, more, move, noble, person, pile, plod, pole, pool, probe, prose, prove, proven, proving, ring, robe, room, sing, sling, slip, solemn, spring, vine

Mathemactivities
active, activities, ash, ate, attic, came, cash, catches, caves, cease, chase, chimes, ham, have, heat, hem, his, mash, mat, match, math, meat, same, sect, shame, sit, state, steam, tact, tames, team, them, themes, ties, times, vice

Investigations
gain, gas, gate, gave, give, given, gone, ingest, invent, invest, invite, nation, negate, nest, nine, noise, note, onset, sing, sit, site, snag, son, stain, state, station, sting, stint, tag, ten, tin, tote, vat, vest, vine, vote